# CAMBRIDGE LIBRARY COLLECTION

*Books of enduring scholarly value*

## Darwin

Two hundred years after his birth and 150 years after the publication of 'On the Origin of Species', Charles Darwin and his theories are still the focus of worldwide attention. This series offers not only works by Darwin, but also the writings of his mentors in Cambridge and elsewhere, and a survey of the impassioned scientific, philosophical and theological debates sparked by his 'dangerous idea'.

## The Coming of Evolution

John Wesley Judd (1840-1916) had a distinguished career, serving as both President of the Geological Society and Dean of the Royal College of Science. Before his retirement as Professor of Geology from Imperial College, he wrote this concise and accessible review of the beginnings of evolutionary theory. Judd skilfully examined the roots of an idea that, already by 1910, had profoundly influenced every branch of science and permeated the work of historians, politicians and theologians. His lively narrative introduces the key individuals, including Darwin and Lyell, who brought about this intellectual revolution. Judd analyses the principal influences that worked upon these scientists as well as the factors that permitted them to remain open to radical new views. His appreciation of the vision, courage and far-reaching impact of the work of both Lyell and Darwin, and the interplay between their ideas, is persuasively and eloquently expressed.

Cambridge University Press has long been a pioneer in the reissuing of out-of-print titles from its own backlist, producing digital reprints of books that are still sought after by scholars and students but could not be reprinted economically using traditional technology. The Cambridge Library Collection extends this activity to a wider range of books which are still of importance to researchers and professionals, either for the source material they contain, or as landmarks in the history of their academic discipline.

Drawing from the world-renowned collections in the Cambridge University Library, and guided by the advice of experts in each subject area, Cambridge University Press is using state-of-the-art scanning machines in its own Printing House to capture the content of each book selected for inclusion. The files are processed to give a consistently clear, crisp image, and the books finished to the high quality standard for which the Press is recognised around the world. The latest print-on-demand technology ensures that the books will remain available indefinitely, and that orders for single or multiple copies can quickly be supplied.

The Cambridge Library Collection will bring back to life books of enduring scholarly value across a wide range of disciplines in the humanities and social sciences and in science and technology.

# The Coming of
# Evolution

*The Story of a Great Revolution in Science*

JOHN W. JUDD

CAMBRIDGE UNIVERSITY PRESS

Cambridge New York Melbourne Madrid Cape Town Singapore São Paolo Delhi

Published in the United States of America by Cambridge University Press, New York

www.cambridge.org
Information on this title: www.cambridge.org/9781108004367

© in this compilation Cambridge University Press 2009

This edition first published 1910
This digitally printed version 2009

ISBN 978-1-108-00436-7

The Cambridge Manuals of Science and
Literature

# THE COMING OF EVOLUTION

CAMBRIDGE UNIVERSITY PRESS
London: FETTER LANE, E.C.
C. F. CLAY, Manager

Edinburgh: 100, PRINCES STREET
London: H. K. LEWIS, 136, GOWER STREET, W.C.
Berlin: A. ASHER AND CO.
Leipzig: F. A. BROCKHAUS
New York: G. P. PUTNAM'S SONS
Bombay and Calcutta: MACMILLAN AND CO., Ltd.

# THE COMING
# OF EVOLUTION

## THE STORY OF A GREAT
## REVOLUTION IN SCIENCE

by

JOHN W. JUDD

C.B., LL.D., F.R.S.

Formerly Professor of Geology and
Dean of the Royal College of Science

Cambridge :
at the University Press
1910

𝕮𝖆𝖒𝖇𝖗𝖎𝖉𝖌𝖊:

PRINTED BY JOHN CLAY, M.A.

AT THE UNIVERSITY PRESS

*With the exception of the coat of arms at the foot, the design on the title page is a reproduction of one used by the earliest known Cambridge printer, John Siberch, 1521*

# CONTENTS

## PLATES

# CHAPTER I

## INTRODUCTORY

WHEN the history of the Nineteenth Century—
'the Wonderful Century,' as it has, not inaptly, been
called—comes to be written, a foremost place must
be assigned to that great movement by which evolution has become the dominant factor in scientific
progress, while its influence has been felt in every
sphere of human speculation and effort. At the
beginning of the Century, the few who ventured
to entertain evolutionary ideas were regarded by
their scientific contemporaries, as wild visionaries
or harmless 'cranks'—by the world at large, as
ignorant 'quacks' or 'designing atheists.' At the
end of the Century, evolution had not only become
the guiding principle of naturalists, but had profoundly influenced every branch of physical science ;
at the same time, suggesting new trains of thought
and permeating the language of philologists, historians, sociologists, politicians—and even of theologians.

How has this revolution in thought—the greatest
which has occurred in modern times—been brought

about? What manner of men were they who were the leaders in this great movement? What the influences that led them to discard the old views and adopt new ones? And, under what circumstances were they able to produce the works which so profoundly affected the opinions of the day? These are the questions with which I propose to deal in the following pages.

It has been my own rare good fortune to have enjoyed the friendship of all the great leaders in this important movement—of Huxley, Hooker, Scrope, Wallace, Lyell and Darwin—and, with some of them, I was long on terms of affectionate intimacy. From their own lips I have learned of incidents, and listened to anecdotes, bearing on the events of a memorable past. Would that I could hope to bring before my readers, in all their nobility, a vivid picture of the characteristics of the men to whom science and the world owe so much!

For it is not only by their intellectual greatness that we are impressed. Every man of science is proud, and justly proud, of the grandeur of character, the unexampled generosity, the modesty and simplicity which distinguished these pioneers in a great cause. It is unfortunately true, that the votaries of science—like the cultivators of art and literature—have sometimes so far forgotten their high vocation, as to have been more careful about the priority

of their personal claims than of the purity of their
own motives—they have sometimes, it must be sadly
admitted, allowed self-interest to obscure the interests
of science. But in the story we have to relate there
are no 'regrettable incidents' to be deplored ; never
has there occurred any event that marred the harmony
in this band of fellow-workers, striving towards a
great ideal. So noble, indeed, was the great central
figure—Charles Darwin—that his senior Lyell and
all his juniors were bound to him by the strongest
ties of admiration, respect and affection ; while he,
in his graceful modesty, thought more of them than
of himself, of the results of their labours rather than
of his own great achievement.

It is not, as sometimes suggested, the striking out
of new ideas which is of the greatest importance in
the history of science, but rather the accumulation
of observations and experiments, the reasonings
based upon these, and the writings in which facts
and reasonings are presented to the world—by which
a merely suggestive hypothesis becomes a vivifying
theory—that really count in making history.

Talking with Matthew Arnold in 1871, he laughingly remarked to me 'I cannot understand why you
scientific people make such a fuss about Darwin.
Why it's all in Lucretius!' On my replying, 'Yes!
Lucretius guessed what Darwin proved,' he mischievously rejoined 'Ah! that only shows how much

greater Lucretius really was,—for he divined a truth,
which Darwin spent a life of labour in groping for.'

Mr Alfred Russel Wallace has so well and clearly
set forth the essential difference between the points
of view of the cultivators of literature and science
in this matter, that I cannot do better than to quote
his words.   They are as follows :—

'I have long since come to see that no one deserves either
praise or blame for the *ideas* that come to him, but only for the
*actions* resulting therefrom.   Ideas and beliefs are certainly not
voluntary acts.   They come to us—we hardly know *how* or
*whence*, and once they have got possession of us we cannot reject
them or change them at will.   It is for the common good that the
promulgation of ideas should be free—uninfluenced by either
praise or blame, reward or punishment.'

'But the *actions* which result from our ideas may properly be so
treated, because it is only by patient thought and work that new
ideas, if good and true, become adopted and utilized; while,
if untrue or if not adequately presented to the world, they are
rejected or forgotten[1].'*

*Ideas* of Evolution, both in the Organic and the
Inorganic world, existed but remained barren for
thousands of years.   Yet by the labours of a band
of workers in last century, these ideas, which were
but the dreams of poets and the guesses of philosophers, came to be the accepted creed of working
naturalists, while they have profoundly affected
thought and language in every branch of human
enterprise.

* For References see the end of the volume.

# CHAPTER II

## ORIGIN OF THE IDEA OF EVOLUTION

IN all ages, and in all parts of the world, we find that primitive man has delighted in speculating on the birth of the world in which he lives, on the origin of the living things that surround him, and especially on the beginnings of the race of beings to which he himself belongs. In a recent very interesting essay[2], the author of *The Golden Bough* has collected, from the records of tradition, history and travel, a valuable mass of evidence concerning the legends which have grown out of these speculations. Myths of this kind would appear to fall into two categories, each of which may not improbably be associated with the different pursuits followed by the uncivilised races of mankind.

Tillers of the soil, impressed as they must have been by the great annual miracle of the outburst of vegetable life as spring returns, naturally adopted one of these lines of speculation. From the dead,

bare ground they witnessed the upspringing of all
the wondrous beauty of the plant-world, and, in their
ignorance of the chemistry of vegetable life, they
imagined that the herbs, shrubs and trees are all
alike built up out of the materials contained in the
soil from which they grow.   The recognition of the
fact that animals feed on plants, or on one another,
led to the obvious conclusion that the *ultimate*
materials of animal, as well as of vegetable, structures
were to be sought for in the soil.   And this view was
confirmed by the fact that, when life ceases in plants
or animals, all alike are reduced to 'dust' and again
become a part of the soil—returning 'earth to earth.'
In groping therefore for an explanation of the origin
of living things, what could be more natural than the
supposition that the first plants and animals—like
those now surrounding us—were made and fashioned
from the soil, dust or earth—all had been 'clay in
the hands of a potter.'   The widely diffused notion
that man himself must have been moulded out of *red*
clay is probably accounted for by the colour of our
internal organs.

Thus originated a large class of legendary stories,
many of them of a very grotesque character.   Even
in many mediaeval sculptures, in this country and on
the continent, the Deity is represented as moulding
with his hands the semblance of a human figure out
of a shapeless lump of clay.

But among the primitive hunters and herdsmen
a very different line of speculation appears to have
originated, for by their occupations they were continually brought into contact with an entirely different
class of phenomena. They could not but notice that
the creatures which they hunted or tended, and slew,
presented marked resemblances to themselves—in
their structures, their functions, their diseases, their
dispositions, and their habits. When dogs and horses
became the servants and companions of men, and
when various beasts and birds came to be kept as
pets, the mental and even the moral processes
characterising the intelligence of these animals must
have been seen by their masters to be identical in
kind with those of their own minds. Do we not even
at the present day compare human characteristics
with those of animals, the courage of the lion, the
cunning of the fox, the fidelity of the dog, and the
parental affection of the bird? And the men, who
depended for their very existence on studying the
ways of various animals, could not have been less
impressed by these qualities than are we.

Mr Frazer has shown how, from such considerations, the legends concerning the relations of certain
tribes of men with particular species of animals have
arisen, and thus the cults of 'sacred animals' and of
'totemism' have been gradually developed. From
comparisons of human courage, sagacity, swiftness,

strength or perseverance, with similar qualities displayed by certain animals, it was an easy transition
to the idea that such characteristics were derived by
inheritance.

In the absence of any exact knowledge of anatomy
and physiology, the resemblances of animals to
themselves would quite outbulk the differences in
the eyes of primitive men, and the idea of close
relationship in blood does not appear to have been
regarded with distaste.   In their origin and in their
destiny, no distinction was drawn between man and
what we now  designate  as  the 'lower' animals.
Primitive man not only feels no repugnance to such
kinship :—

> 'But thinks, admitted to that equal sky,
> His faithful dog shall bear him company[3].'

It should perhaps be remembered, too, that, in
the breeding of domestic animals, the great facts of
heredity and variation could not fail to have been
noticed, and must have given rise to reflection and
speculation.  The selection of the best animals for
breeding purposes, and the consequent improvement
of their stock, may well have suggested the transmutation of one kind of animal into a different kind,
just as the crossing of different kinds of animals
seems to have suggested the possible existence of
centaurs, griffins and other monstrous forms.

How early the principles of variation and heredity, and even the possibility of improving breeds by selection, must have been appreciated by early men is illustrated by the old story of the way in which the wily Jacob made an attempt—however futile were the means he adopted—to cheat his employer Laban[4].

Yet, in spite of observed tendencies to variation among animals and plants, early man must have been convinced of the existence of distinct kinds ('species') in both the vegetable and animal worlds; he recognised that plants of definite kinds yielded particular fruits, and that different kinds of animals did not breed promiscuously with one another, but that, pairing each with its own kind, all gave rise to like offspring, and thus arose the idea of distinct 'species' of plants and animals.

It must be remembered, however, that for a long time 'the world' was believed to be limited to a few districts surrounding the Eastern Mediterranean, and the kinds or 'species' of animals and plants were supposed to number a few scores or at most hundreds. This being the case, the sudden stocking of 'the world' with its complement of animals and plants would be thought a comparatively simple operation, and the violent destruction of the whole a scarcely serious result. Even the possibility of the preservation of pairs of all the different species, in a ship of moderate dimensions, was one that was easily enter-

tained and was not calculated to awaken either sur-
prise or incredulity.

But how different is the problem as it now presents
itself to us ! In the year 1900 Professor S. H. Vines
of Oxford estimated that the number of 'species' of
plants that have been described could be little short
of 200,000, and that future studies, especially of the
lower microscopic forms, would probably bring that
number up to 300,000[5]. Last year, Mr A. E. Shipley
of Cambridge, basing his estimate on the earlier one
of Dr Günther, came to the conclusion that the number
of described animals must also exceed 300,000[6]. On
the lowest estimate then we must place the number
of known species of plants and animals, living on the
globe, as 600,000 ! And if we consider the numbers
of new forms of plants and animals that every year
are being described by naturalists—about 1500 plants
and 1200 animals—if we take into account the inac-
cessible or as yet unvisited portions of the earth's
surface, the very imperfectly known depths of the sea,
and, in addition to these, the almost infinite varieties
of minute and microscopic forms, I think every competent judge would consider *a million* as being
probably an estimate below, rather than above, the
number of 'species' now existing on the earth !

While some of these species are very widely
distributed over the earth's surface, or in the waters
of the oceans, seas, lakes and rivers, there are others

which are as strikingly limited in their range.  Many
of the myriad forms of insect-life pass their whole
existence, and are dependent for food, on a particular
species of plant.  Not a few animals and plants are
parasitical, and can only live in the interior or on the
outside of other plants and animals.

It will be seen from these considerations that in
attempting to decide between the two hypotheses of
the *origin* of species—the only ones ever suggested—
namely the fashioning of them out of dead matter, or
their descent with modification from pre-existing
forms, we are dealing with a problem of much greater
complexity than could possibly have been imagined
by the early speculators on the subject.

The two strongly contrasted hypotheses to which
we have referred are often spoken of as 'creation'
and 'evolution.'  But this is an altogether illegitimate
use of these terms.  By *whatever method* species of
plants or animals come into existence, they may be
rightly said to be 'created.'  We speak of the
existing plants and animals as having been created,
although we well know them to have been 'evolved'
from seeds, eggs and other 'germs'—and indeed from
those excessively minute and simple structures known
as 'cells.'  Lyell and Darwin, as we shall presently
see, though they were firmly convinced that species of
plants and animals were slowly developed and not
suddenly manufactured, wrote constantly and correctly
of the 'creation' of new forms of life.

The idea of 'descent with modification,' derived
from the early speculations of hunters and herdsmen,
is really a much nobler and more beautiful conception
of 'creation' than that of the 'fashioning out of
clay,' which commended itself to the primitive agriculturalists.

Lyell writing to his friend John Herschel, who
like himself believed in the derivation of new species
from pre-existing ones by the action of secondary
causes, wrote in 1836 :—

When I first came to the notion,...of a succession of
extinction of species, and creation of new ones, going on per-
petually now, and through an indefinite period of the past, and to
continue for ages to come, all in accommodation to the changes
which must continue in the inanimate and habitable earth, the
idea struck me as the grandest which I had ever conceived, so far
as regards the attributes of the Presiding Mind[7].'

And Darwin concludes his presentment of the
doctrine of evolution in the *Origin of Species* in 1859
with the following sentence :—

'There is a grandeur in this view of life, with its several
powers, having been originally breathed by the Creator into a few
forms or into one ; and that, whilst this planet has gone cycling on
according to the fixed law of gravity, from so simple a beginning
endless forms most beautiful and most wonderful have been, and
are being, evolved[8].'

Compare with these suggestions the ideas embodied in the following lines—ideas of which the

crudeness cannot be concealed by all the witchery of
Milton's immortal verse :—

> 'The Earth obey'd, and straight,
> Op'ning her fertile womb, teem'd at a birth
> Innumerous living creatures, perfect forms,
> Limb'd and full grown.  Out of the ground up rose
> As from his lair, the wild beast, where he wons
> In forest wild, in thicket, brake, or den ;
> Among the trees they rose, they walk'd ;
> The cattle in the fields and meadows green :
> Those rare and solitary, these in flocks
> Pasturing at once, and in broad herds upsprung.
> The grassy clods now calv'd ; now half appear'd
> The tawny lion, pawing to get free
> His hinder parts, then springs, as broke from bonds,
> And rampant shakes his brinded mane[9].'

Can anyone doubt for a moment which is the
grander view of ' Creation '—that embodied in
Darwin's prose, or the one so strikingly pictured in
Milton's poetry ?

We see then that the two ideas of the method of
creation, dimly perceived by early man, have at last
found clear and definite expression from these two
authors—Milton and Darwin.  It is a singular coincidence that these two great exponents of the rival
hypotheses were both students in the same University
of Cambridge and indeed resided in the same foundation—and that not one of the largest of that
University—namely Christ's College.

# CHAPTER III

## THE DEVELOPMENT OF THE IDEA OF EVOLUTION TO THE INORGANIC WORLD

WE have seen in the preceding chapter that, with respect to the origin of plants and animals—including man himself—two very distinct lines of speculation have arisen; these two lines of thought may be expressed by the terms 'manufacture'—literally making by hand, and 'development' or 'evolution,' —a gradual unfolding from simpler to more complex forms. Now with respect to the *inorganic* world two parallel hypotheses of 'creation' have arisen, like those relating to *organic* nature; but in the former case the determining factor in the choice of ideas has been, not the avocations of the primitive peoples, but the nature of their surroundings.

The dwellers in the valleys of the Euphrates and Tigris could not but be impressed by the great and destructive floods to which those regions were subject; and the inhabitants of the shores and islands of the Aegean Sea, and of the Italian peninsula, were equally conversant with the devastations wrought by volcanic outbursts and earthquake shocks.  As great districts were seen to be depopulated by these catastrophies, might not some even more violent cataclysm of the same kind actually destroy all mankind, with the animals and plants, in the comparatively small area then known as 'the world'?  The great flood, of which all these nations appear to have retained traditions, was regarded as only the last of such destructive cataclysms ; and, in this way, there originated the myth of successive destructions of the face of the earth, each followed by the creation of new stocks of plants and animals.  This is the doctrine now known as 'Catastrophism,' which we find prevalent in the earliest traditions and writings of India, Babylonia, Syria and Greece.

But in ancient Egypt quite another class of phenomena was conspicuously presented to the early philosophers of the country.  Instead of sudden floods and terrible displays of volcanic and earthquake violence, they witnessed the annual gentle rise and overflowings of their grand river, with its beneficent heritage of new soil ; and they soon learned to recognise that Egypt itself—so far as the delta was concerned—was 'the gift of the Nile.'

From the contemplation of these phenomena, the Egyptian sages were gradually led to entertain the idea that all the features of the earth—as they knew

it—might have been similarly produced through the
slow and constant action of the causes now seen in
operation around them. This idea was incorporated
in a myth, which was suggested by the slow and
gradual transformation of an egg into a perfect,
growing organism. The birth of the world was
pictured as an act of incubation, and male and female
deities were invented to play the part of parents to
the infant world. By Pythagoras, who resided for
more than twenty years in Egypt, these ideas were
introduced to the Greek philosophers, and from that
time 'Catastrophism' found a rival in the new
doctrine which we shall see has been designated under
the names of 'Continuity,' 'Uniformitarianism' or
'Evolution.' How, from the first crude notions of
evolution, successive thinkers developed more just and
noble conceptions on the subject, has been admirably
shown by Professor Osborn in his *From the Greeks to
Darwin* and by Mr Clodd in his *Pioneers of Evolution.*

Poets, from Empedocles and Lucretius to Goethe
and Tennyson, have sought in their verses to illustrate
the beauty of evolutionary ideas ; and philosophers,
from Aristotle and Strabo to Kant and Herbert
Spencer, have recognised the principle of evolution
as harmonising with, and growing out of, the highest
conceptions of science. Yet it was not till the Nineteenth Century that any serious attempts were made
to establish the hypothesis of evolution as a definite

theory, based on sound reasoning from careful observation.

It is true that there were men, in advance of their age, who in some cases anticipated to a certain extent this work of establishing the doctrine of evolution on a firm foundation.  Thus in Italy, the earliest home of so many sciences, a Carmelite friar, Generelli, reasoning on observations made by his compatriots Fracastoro and Leonardo da Vinci in the Sixteenth Century, Steno and Scilla in the Seventeenth, and Lazzaro Moro and Marsilli in the Eighteenth Century, laid the foundations of a rational system of geology in a work published in 1749 which was characterised alike by courage and eloquence.  In France, the illustrious Nicolas Desmarest, from his study of the classical region of the Auvergne, was able to show, in 1777, how the river valleys of that district had been carved out by the rivers that flow in them.  Nor were there wanting geologists with similar previsions in Germany and Switzerland.

But none of these early exponents of geological theory came so near to anticipating the work of the Nineteenth Century as did the illustrious James Hutton, whose 'Theory of the Earth,' a first sketch of which was published in 1785, was a splendid exposition of evolution as applied to the inorganic world. Unfortunately, Hutton's theory was linked to the extravagancies of what was known at that day as

'Vulcanism' or 'Plutonism,' in contradistinction to
the 'Neptunism' of Werner. Hutton, while rejecting
the Wernerian notion of " the aqueous precipitation
of basalt," maintained the equally fanciful idea that
the consolidation of all strata—clays, sandstones,
conglomerates, limestones and even rock-salt—must
be ascribed to the action of heat, and that even the
formation of chalk-flints and the silicification of fossil
wood were due to the injection of molten silica!

What was still more unfortunate in Hutton's case
was that, in his enthusiasm, he used expressions which
led to his being charged with heresy and even with
being an enemy of religion. His writings were
further so obscure in style as often to lead to misconception as to their true meaning, while his great work
—so far as the fragment which was published goes—
contained few records of original observations on
which his theory was based.

Dr Fitton has pointed out very striking coinci-
dences between the writings of Generelli and those of
Hutton, and has suggested that the latter may have
derived his views from the eloquent Italian friar [10].
But for this suggestion, I think that there is no real
foundation. Darwin and Wallace, as we shall see
later, were quite unconscious of their having been
forestalled in the theory of Natural Selection by
Dr Wells and Patrick Matthew; and Hutton, like
his successor Lyell, in all probability arrived, quite

independently, and by different lines of reasoning, at conclusions identical with those of Generelli and Desmarest.

Although, as we shall see, Hutton failed to greatly influence the scientific thought of his day, yet all will now agree with Lyell that 'Hutton laboured to give fixed principles to geology, as Newton had succeeded in doing to astronomy[11]'; and with Zittel that *Hutton's Theory of the Earth* is one of the master-pieces in the history of geology[12].'

# CHAPTER IV

## THE TRIUMPH OF CATASTROPHISM OVER
## EVOLUTION

THERE is no fact in the history of science which is more certain than that those great pioneers of Evolution in the Inorganic world—Generelli, Desmarest and Hutton—utterly failed to recommend their doctrines to general acceptance ; and that, at the beginning of last century, everything in the nature of evolutionary ideas was almost universally discredited —alike by men of science and the world at large.

The causes of the neglect and opprobrium which befel all evolutionary teachings are not difficult to discover. The old Greek philosophers saw no more reason to doubt the possibility of creation by evolution, than by direct mechanical means. But, on the revival of learning in Europe, evolution was at once confronted by the cosmogonies of Jewish and Arabian writers, which were incorporated in sacred books ; and not only were the ideas of the sudden making and destruction of the world and all things in it regarded as revealed truth, but the periods of time necessary for evolution could not be admitted by those who believed the beginning of the world to have been recent, and its end to be imminent. Thus 'Catastrophic' ideas came to be regarded as *orthodox*, and evolutionary ones as utterly irreligious and damnable.

There are few more curious facts in the history of science than the contrast between the reception of the teaching of the Saxon professor Werner, and those of Hutton, the Scotch philosopher, his great rival. While the enthusiastic disciples of the former carried their master's ideas everywhere, acting with missionary zeal and fervour, and teaching his doctrines almost as though they were a divine revelation, the latter, surrounded by a few devoted friends, saw his teachings everywhere received with persistent misrepresentation, theological vituperation or contemptuous neglect. Even in Edinburgh itself, one of Werner's pupils dominated the teaching of the University for half a century, and established a society for the propagation of the views which Hutton so strongly opposed.

When it is remembered that Hutton wrote at a time when 'heresy-hunting' in this country had been excited to such a dangerous extent, through the excesses of the French Revolution, that his contemporary, Priestley, had been hounded from his home and country for proclaiming views which at that

time were regarded as unscriptural, it becomes less
difficult to understand the prejudice that was excited
against the gentle and modest philosopher of
Edinburgh.

We have employed the term 'Catastrophism' to
indicate the views which were prevalent at the
beginning of last century concerning the origin of the
rock-masses of the globe and their fossil contents.
These views were that at a number of successive
epochs—of which the age of Noah was the latest—
great revolutions had taken place on the earth's
surface ; that during each of these cataclysms all
living things were destroyed ; and that, after an
interval, the world was restocked with fresh assem-
blages of plants and animals, to be destroyed in turn
and entombed in the strata at the next revolution.

Whewell, in 1830, contrasted this teaching with
that of Hutton and Lyell in the following passage :—
'These two opinions will probably for some time
divide the geological world into two sects, which may
perhaps be designated the "Uniformitarians" and
the "Catastrophists." The latter has undoubtedly
been of late the prevalent doctrine.' It is interesting
to note, as showing the confidence felt in their tenets
by the 'Catastrophists' of that day, that Whewell
adds 'We conceive that Mr Lyell will find it a harder
task than he imagines to overturn the established
belief[13] ! '

Some authors have suggested that the doctrine taught by Generelli, Desmarest and Hutton, and later by Scrope and Lyell, for which Whewell proposed the somewhat cumbrous term 'Uniformitarianism,' but which was perhaps better designated by Grove in 1866 as 'Continuity[14],' was distinct from, and subsidiary to, Evolution—and this view could claim for a time the support of a very great authority.

In 1869, Huxley delivered an address to the Geological Society, in which he postulated the existence of 'three more or less contradictory systems of geological thought,' under the names of 'Catastrophism,' 'Uniformitarianism' and 'Evolution.' In this essay, distinguished by all his wonderful lucidity and forceful logic, Huxley sought to establish the position that evolution is a doctrine, distinct from and *in advance of* that of uniformitarianism, and that Hutton and Playfair—'and to a less extent Lyell'—had acted unwisely in deprecating the extension of Geology into enquiries concerning 'the beginning of things[15].'

But there is no doubt that Huxley at a later period was led to qualify, and indeed to largely modify, the views maintained in that address. In a footnote to an essay written in April 1887, he asserts 'What I mean by "evolutionism" is consistent and thoroughgoing uniformitarianism'; and in the same year he wrote in his *Reception of the Origin of*

*Species*[16]: 'Consistent uniformitarianism postulates evolution, as much in the organic as in the inorganic world[17].'

It is not difficult to trace the causes of this change in the attitude of mind with which Huxley regarded the doctrine of 'uniformitarianism.' He assures us 'I owe more than I can tell to the careful study of the *Principles of Geology*[18],' and again 'Lyell was for others as for me the chief agent in smoothing the road for Darwin[19].' From the perusal of the letters of Lyell, published in 1881, Huxley learned that the author of the *Principles of Geology* had, at a very early date, been convinced that evolution was true of the organic as well as of the inorganic world—though he had been unable to accept Lamarckism, or any other hypothesis on the subject that had, up to that time, been suggested. There can be little doubt, however, that a chief influence in bringing about the change in Huxley's views was his intercourse with Darwin—who was, from first to last, an uncompromising 'uniformitarian.'

We are fully justified, then, in regarding the teaching of Hutton and Lyell (to which Whewell gave the name of 'uniformitarianism' as being identical with evolution. The cockpit in which the great battle between catastrophism and evolution was fought out, as we shall see in the sequel, was the Geological Society of London, where doughty champions of each of the rival doctrines met in frequent combat and long maintained the struggle for supremacy.

Fitton has very truly said that 'the views proposed by Hutton failed to produce general conviction at the time ; and several years elapsed before any one showed himself publicly concerned about them, either as an enemy or a friend[20].' Sad is it to relate that, when notice was at last taken of the memoir on the 'Theory of the Earth,' it was by bitter opponents —such 'Philistines' (as Huxley calls them) as Kirwan, De Luc and Williams, who declared the author to be an enemy of religion. Not only did Hutton, unlike the writers of other theories of the earth, omit any statement that his views were based on the Scriptures, but, carried away by the beauty of the system of continuity which he advocated, he wrote enthusiastically 'the result of this physical enquiry is that we find no vestige of a beginning—no prospect of an end[21].' This was unjustly asserted to be equivalent to a declaration that the world had neither beginning nor end ; and thus it came about that Wernerism, Neptunism and Catastrophism were long regarded as synonymous with Orthodoxy, while Plutonism and 'Uniformitarianism' were looked upon with aversion and horror as subversive of religion and morality.

Almost simultaneously with the foundation of the Wernerian Society of Edinburgh (in 1807) was the

establishment in London of the Geological Society.
Originating in a dining club of collectors of minerals,
the society consisted at first almost exclusively of
mineralogists and chemists, including Davy, Wollaston,
Sir James Hall, and later, Faraday and Turner. The
bitter but barren conflict between the Neptunists and
the Plutonists was then at its height, and it was, from
the first, agreed in the infant society to confine its
work almost entirely to the collection of facts,
eschewing theory. During the first decade of its
existence, it is true, the chief papers published by
the society were on mineralogical questions; but
gradually geology began to assert itself. The actual
founder and first president of the society, Greenough,
had been a pupil of Werner, and used all his great
influence to discourage the dissemination of any but
Wernerian doctrines—foreign geologists, like Dr
Berger, being subsidised to apply the Wernerian
classification and principles to the study of British
rocks. Thus, in early days, the Geological Society
became almost as completely devoted to the teaching
of Wernerian doctrines as was the contemporary
society in Edinburgh.

Dr Buckland used to say that when he joined the
Geological Society in 1813, 'it had a very *landed*
manner, and only admitted the professors of geology
in Oxford and Cambridge on sufferance.'

But, gradually, changes began to be felt in this

aristocratic body of exclusive amateurs and wealthy collectors of minerals. William Smith, 'the Father of English Geology'—though he published little and never joined the society—exercised a most important influence on its work. By his maps, and museum of specimens, as well as by his communications, so freely made known, concerning his method of 'identifying strata by their organic remains,' many of the old geologists, who were not aware at the time of the source of their inspiration, were led to adopt entirely new methods of studying the rocks. In this way, the accurate mineralogical and geognostical methods of Werner came to be supplemented by the fruitful labours of the stratigraphical palaeontologist. The new school of geologists included men like William Phillips, Conybeare, Sedgwick, Buckland, De la Beche, Fitton, Mantell, Webster, Lonsdale, Murchison, John Phillips and others, who laid the foundations of British stratigraphical geology.

But these great geological pioneers, almost without exception, maintained the Wernerian doctrines and were firm adherents of Catastrophism. The three great leaders—the enthusiastic Buckland, the eloquent Sedgwick, and the indefatigable Conybeare—were clergymen, as were also Whewell and Henslow, and they were all honestly, if mistakenly, convinced that the Huttonian teaching was opposed to the Scriptures

and inimical to religion and morality. Buckland at
Oxford, and Sedgwick at Cambridge, made geology
popular by combining it with equestrian exercise;
and Whewell tells us how the eccentric Buckland used
to ride forth from the University, with a long caval-
cade of mounted students, holding forth with sarcasm
and ridicule concerning 'the inadequacy of existing
causes[22].'

And Sedgwick at Cambridge was no less firmly
opposed to evolutionary doctrine, eloquently declaim-
ing at all times against the unscriptural tenets of the
Huttonians.

I cannot better illustrate the complete neglect at
that time by leading geologists in this country of the
Huttonian teaching than by pointing to the Report
drawn up in 1833, by Conybeare, for the British
Association, on 'The Progress, Actual State and
Ulterior Prospects of Geological Science[23].' This
valuable memoir of 47 pages opens with a sketch of
the history of the science, in which the chief Italian,
French and German investigators are referred to, but
the name of Hutton is not even mentioned!

And if positive evidence is required of the con-
tempt which the early geologists felt for Hutton and
his teachings, it will be found in the same author's
introduction to that classical work, the *Outlines of
Geology* (1822), in which he says of Hutton, after

praising his views on granite veins and "trap
rocks " :—

'The wildness of many of his theoretical views, however, went
far to counterbalance the utility of the additional facts which he
collected from observation. He who could perceive in geology
nothing but the *ordinary* operation of actual causes, carried
on in the same manner through infinite ages, without the
trace of a beginning or the prospect of an end, must have
surveyed them through the medium of a preconceived hypothesis
alone[24].'

John Playfair, the brilliant author of the *Illustrations of the Huttonian Theory*, died in 1819 ; under
happier conditions his able work might have done for
Inorganic Evolution what his great master failed to
accomplish ; but the dead weight of prejudice and the
dread of anything that seemed to savour of infidelity
was, at the time of the great European struggle
against revolutionary France, too great to be removed
even by his lucid statements and eloquent advocacy.
James Hall and Leonard Horner, two faithful disciples
of Hutton, who had joined the infant Geological
Society, forsook it early, the former leaving it on
account of the quarrel with the Royal Society, the
latter retaining his fellowship and interest, but going
to live at Edinburgh. Greenough, 'The Objector
General,' as he was called, was left, fanatically
opposing any attempt to stem the current that had

set so strongly in favour of Wernerism and Neptunism, and the Catastrophic doctrines which all thought to be necessary conclusions from them. The great heroic workers of that day—while they were laying well and truly the foundations of historical geology—were, one and all, indifferent to, or violently opposed to, the Huttonian teaching. Neither Fitton nor John Phillips, who at a later date showed sympathy with evolutionary doctrines, were the men to fight the battle of an unpopular cause.

Attempts have been made by both Playfair and Fitton to explain how it was that Hutton's teaching failed to arrest the attention it deserved. The former justly asserted that the world was tired of the performances issued under the title of 'theories of the earth'; and that the condensed nature of Hutton's writings, with their 'embarrassment of reasoning and obscurity of style [25]' are largely responsible for the neglect into which they fell.

Fitton, in 1839, wrote in the *Edinburgh Review*, 'The original work of Hutton (in two volumes) is in fact so scarce that no very great number of our readers can have seen it. No copy exists at present in the libraries of the Royal Society, the Linnean, or even the Geological Society of London [26] !' He also points out that Hutton's work, and even the more lucid *Illustrations of the Huttonian Theory*, were almost unknown on the continent, owing to the isolation of Great Britain during the war ; and he even suggests that the popularity of Playfair in this country may have not improbably led to the neglect of the original work of Hutton[27].

On the continent, indeed, the authority of Cuvier was supreme, and in his *Essay on the Theory of the Earth*, prefixed to his *Opus magnum*—the *Ossemens Fossiles*—the great naturalist threw the whole weight of his influence into the scale of Catastrophism. He maintained that a series of tremendous cataclysms had affected the globe—the last being the Noachian deluge—and that the floods of water that overspread the earth, during each of these events, had buried the various groups of animals, now extinct, that had been successively created.

If anything had been wanted in England to support and confirm the views that were then supposed to be the only ones in harmony with the Scriptures, it was found in the great authority of Cuvier. As Zittel justly says, Cuvier's theory of 'World-Cata-strophies'—'which afforded a certain scientific basis for the Mosaic account of the "Flood," was received with special cordiality in England, for there, more than in any other country, theological doctrines had always affected geological conceptions [28].' Britain, which had produced the great philosopher, Hutton, had now become the centre of the bitterest opposition to his teachings !

But 'the darkest hour of night is that which precedes the dawn,' and while the forces of reaction in this country appeared to be triumphant over Hutton's teaching, there was in preparation, to use the words of Darwin, a 'grand work'...'which the future historian will recognise as having produced a revolution in natural science.'

# CHAPTER V

## THE REVOLT OF SCROPE AND LYELL AGAINST CATASTROPHISM

THE year 1797, in which the illustrious Hutton died, leaving behind him the noble fragments of a monumental work, was signalised by the birth of two men, who were destined to bring about the overthrow of Catastrophism, and to establish, upon the firm foundation of reasoned observation, the despised doctrine of Uniformitarianism or Evolution —as outlined by Generelli, Desmarest and Hutton. These two men were George Poulett Thomson (who afterwards took the name of Scrope) and Charles Lyell. Both of them were, from their youth upwards, brought under the strongest influences of the prevalent anti-evolutionary teachings ; but both emancipated themselves from the effects of these teachings, being led gradually by their geological travels and observations, not only to reject their early faith, but to become the champions of Evolution.

There was a singular parallel between the early
careers of these two men. Both were the sons of
parents of ample means, and were thus freed from
the distractions of a business or profession, while
throughout life they alike remained exempt from
family cares. Each of them received the ordinary
education of the English upper classes—Scrope at
Harrow, and Lyell at Salisbury, in a school conducted
by a Winchester master on public-school lines. In
due course, the two young men proceeded to the
University—Scrope to Cambridge, to come under the
influence of the sagacious and eloquent Sedgwick,
and Lyell to Oxford, to catch inspiration from the
enthusiastic but eccentric Buckland. On the opening
up of the continent, by the termination of the French
wars, each of the young men accompanied his family
in a carriage-tour (as was the fashion of the time)
through France, Switzerland and Italy ; and both
utilised the opportunities thus afforded them, to
make long walking excursions for geological study.
They both returned again and again to the continent
for the purpose of geological research, and in the year
1825, at the age of 28, found themselves associated
as joint-secretaries of the Geological Society. By
this time they had arrived at similar convictions
concerning the causes of geological phenomena—
convictions which were in direct opposition to the
views of their early teachers, and equally obnoxious

to all the leaders of geological thought in the infant
society which they had joined.

It is interesting to note that each of these two
young geologists arrived independently, *as the result
of their own studies and observations*, at their
conclusions concerning the futility of the prevailing
catastrophic doctrines. This I am able to affirm, not
only from their published and unpublished letters,
but from frequent conversations I had with them in
their later years.

Scrope, who was slightly the elder of the two
friends, spent a considerable time in that wonderful
district of France—the Auvergne—in the year 1821,
and though he had not seen the map and later
memoirs of Desmarest, he pourtrayed the structure
of the country in a series of very striking panoramic
views, and was led, independently of the great French
observer, to the same conclusions as his concerning
the volcanic origin of the basalts and the formation
of the valleys by river-action. Scrope was at that
time equally ignorant of the views propounded both
by Generelli and by Hutton.

By April 6th, 1822, Scrope had completed his
masterly work *The Geology and Extinct Volcanoes
of Central France*, and had despatched it to England.
It would be idle to speculate now as to what might
have been the effect of that work—so full of the
results of accurate observation, and so suggestive in

its reasoning—had it been published at that time.
It is quite possible that much of the credit now
justly assigned to Lyell, would have belonged to his
friend.  Unfortunately, however, Scrope, instead of
seeing his work through the press, determined first
to make another tour in Italy.  He arrived at Naples
just in time to witness and describe the grandest
eruption of Vesuvius in modern times, that of October
1822.  What he witnessed then—the blowing away
of the whole upper part of the mountain and the
formation of a vast crater 1000 feet deep—made a
profound impression on Scrope's mind.  His interest
thus strongly aroused concerning igneous phenomena,
Scrope continued his travels and observations on the
volcanic rocks of the peninsula of Italy and its
islands, and was thus led to a number of important
conclusions in theoretical geology, which he embodied
in a work, published in 1825, entitled *Considerations
on Volcanos: the probable causes of their phenomena,
the laws which determine their march, the disposition
of their products, and their connexion with the present
state and past history of the globe; leading to the
establishment of a New Theory of the Earth.*

It is only right to point out that, in calling this
book a *new* 'Theory of the Earth,' Scrope had no
intention of comparing it with Hutton's great
work, with which he was at that time altogether
unacquainted.  Nevertheless, his conclusions, though

independently arrived at, were almost identical with
those of the great Scotch philosopher. But Scrope
made the same mistake as Hutton had done before
him. He allowed his theoretical conclusions to
precede, instead of following upon an account of
the observations on which they were based. Scrope's
book is certainly one of the most original and
suggestive contributions ever made to geological
science; but the very speculative character of a
large portion of the work led to the neglect of the
really valuable hypotheses and acute observations
which it contained. In the preface, however, the
author gives a most striking and complete summary
of the doctrine of Evolution as opposed to Catastrophism, in the inorganic world, as will be shown
by the following extracts :—

Geology has for its business a knowledge of the processes
which are in continual or occasional operation within the limits
of our planet, and the application of these laws to explain the
appearances discovered by our Geognostical researches, so as from
these materials to deduce conclusions as to the past history of
the globe.

The surface of the globe exposes to the eye of the Geognost
abundant evidence of a variety of changes which appear to have
succeeded one another during an incalculable lapse of time.

These changes are chiefly,

I. Variations of level between different constituent parts of
the solid surface of the globe.

II. The destruction of former rocks, and their reproduction under another form.

III. The production of rocks *de novo* upon the earth's surface.

Geologists have usually had recourse for the explanation of these changes to the supposition of sundry violent and extraordinary catastrophes, cataclysms, or general revolutions having occurred in the physical state of the earth's surface.

As the idea imparted by the term Cataclysm, Catastrophe, or Revolution, is extremely vague, and may comprehend any thing you choose to imagine, it answers for the time very well as an explanation; that is, it stops further inquiry. But it has also the disadvantage of effectually stopping the advance of science, by involving it in obscurity and confusion.

If, however, in lieu of forming guesses as to what may have been the possible causes and nature of these changes, we pursue that, which I conceive the only legitimate path of geological inquiry, and begin by examining the laws of nature which are actually in force, we cannot but perceive that numerous physical phenomena are going on at this moment on the surface of the globe, by which various changes are produced in its constitution and external characters; changes extremely analogous to those of earlier date, whose nature is the main object of geological inquiry.

These processes are principally,

I. The Atmospheric phenomena.

II. The laws of the circulation and residence of Water on the exterior of the globe.

III. The action of Volcanos and Earthquakes.

The changes effected before our eyes, by the operation of these causes, in the constitution of the crust of the earth are chiefly—

I. The Destruction of Rocks.

II. The Reproduction of others.

III.  Changes of Level.

IV.  The Production of New Rocks from the interior of the globe upon its surface.

Changes which in their general characters bear so strong an analogy to those which are suspected to have occurred in the earlier ages of the world's history, that, until the processes which give rise to them have been maturely studied under every shape, and then applied with strict impartiality to explain the appearances in question ; and until, after a long investigation, and with the most liberal allowances for all possible variations, and an unlimited series of ages, they have been found wholly inadequate to the purpose, it would be the height of absurdity to have recourse to any gratuitous and unexampled hypothesis for the solution of these analogous facts [29].

It was not till 1826, four years after the completion of the work, that Scrope managed to publish his book on the Auvergne, and to tear himself away from the speculative questions by which he had become obsessed. No one could be more candid than he was in acknowledging the causes of his failure to impress his views upon his contemporaries. Writing in 1858, he said of his *Considerations on Volcanos* :—

'In that work unfortunately were included some speculations on theoretic cosmogony, which the public mind was not at that time prepared to entertain. Nor was this my first attempt at authorship, sufficiently well composed, arranged or even printed, to secure a fair appreciation for the really sound and, I believe, original views on many points of geological interest which it contained. I ought, no doubt, to have begun with a description

of the striking facts which I was prepared to produce from the
volcanic regions of Central France and Italy, in order to pave the
way for a favourable reception, or even a fair hearing, of the
theoretical views I had been led from these observations to
form [30].'

He adds that 'this obvious error was pointed out
in a very friendly manner' in a notice of the memoir
on *The Geology of Central France*, which was
contributed by Lyell to the *Quarterly Review* in
1827 [31].

Scrope's geological career however—though one
of so much promise—was brought to a somewhat
abrupt termination. In 1821 he had married the
last representative and heiress of the Scropes, the
old Earls of Wiltshire, and soon afterwards he settled
down at the family seat of Castle Combe, eventually
devoting his attention almost exclusively to social
and political questions. From 1833 to 1868, when
he retired from Parliament, he was member for
Stroud; and though he seldom took part in the
debates, he became famous as a writer of political
tracts, thus acquiring the sobriquet of 'Pamphlet
Scrope.' He himself used to relate an amusing
incident at his own expense. His great friend Lord
Palmerston, on being greeted with the question,
'Have you read my last pamphlet?' replied mischievously, 'Well Scrope, I hope I have!'

It is sad to relate that, owing to a carriage accident,

Scrope's wife became a confirmed invalid and he had
no child to succeed to the estate. Though cut off
by other duties from the geological world, Scrope
maintained his correspondence with his old friend
Lyell, and, as we shall see in the sequel, was able to
render him splendid service by the luminous though
discriminating reviews of the *Principles of Geology*
in the *Quarterly Review*. Throughout his life,
however, Scrope preserved a love of geology, and
occasionally contributed to the literature of the
science ; and in his closing years, when unable to
travel himself, he gave to others the means of carrying on the researches in which he had from the first
been so deeply interested.

Fortunately for science, Lyell's devotion to
geological study was not, like Scrope's, interrupted
by the claims made upon him by social and political
questions. Feeling though he did, with his friend,
the deepest sympathy in all liberal movements, and
being especially interested in the reform of educational methods, his geological work always had the
first claim on his time and attention, and nothing was
allowed to interfere with his scientific labours.

Charles Lyell was the eldest son of a Scottish
laird, whose forbears, after making a fortune in India,
had purchased the estate of Kinnordy in Strathmore,
on the borders of the Highlands. Lyell's father was

a man of culture, a good classical scholar, a translator
and commentator on Dante, and a cryptogamic
botanist of some reputation.

Lyell's mother, an Englishwoman from York-
shire, was a person of great force of character ; this
she showed when, on coming to Kinnordy, she found
drunkenness so prevalent among the lairds of this
part of Scotland, as to cause a fear on her part, that
her husband might be drawn into the dangerous
society : she therefore induced him, when their son
Charles was only three months old, to abandon their
Scottish home, and settle in the New Forest of
Hampshire. Thus it came about that the future
geologist, though born in Scotland, became, by
education, habits and association, English.

Charles Lyell's attention was first drawn to
geology by seeing the quartz-crystals and chalcedony
exposed in the broken chalk-flints, which he, as a boy
of ten, used to roll down, in company with his school-
fellows, from the walls of Old Sarum. Like Charles
Darwin, too, he became an ardent and enthusiastic
collector of insects, and grew to be a tall and active
young fellow, a keen sportsman, with only one draw-
back—a weakness of the eyes which troubled him
through all his after life.

It was when at the age of seventeen he went to
Oxford and came under the influence of Dr Buckland
that Lyell first became deeply engrossed in geology.

Lyell used to tell many amusing stories of the
oddities of his old teacher and friend Buckland.  In
his lectures, both in the University and on public
platforms, Buckland would keep his audience in roars
of laughter, as he imitated what he thought to be
the movements of the iguanodon or megatherium,
or, seizing the ends of his long clerical coat-tails,
would leap about to show how the pterodactyle flew.
Lyell became greatly attached to Buckland, who used
to take him privately on geological expeditions.  On
one of these occasions, they were dining at an inn,
where a gentleman at another table became greatly
scandalised by Buckland's conversation and manners.
The professor, seeing this, became more outrageous
than ever, and on parting with Lyell for the night
took the candle and placed it between his teeth, so
as to illuminate the mouth-cavity exclaiming, 'There
Lyell, practise this long enough and you will be able
to do it as well as I do.'  When Buckland had retired,
the stranger revealed himself to Lyell as an old friend
of his father's, adding 'I hope you will never be seen
in the company of that buffoon again.'  'Oh! Sir,'
said the startled undergraduate, 'that is my professor
at Oxford!'  But Buckland did not always originate
the fun, for Lyell told me that, when the professor
visited Kinnordy in his company, he led him a long
tramp under promise of showing him 'diluvium
intersected by whin dykes,' and, in the end, pointed

to fields in a boulder-clay country separated by gorse ('whin') hedges ('dykes').

Buckland, as shown by his *Vindiciae Geologicae* (1820) and his *Bridgewater Treatise* (1836), was the most uncompromising of the advocates for making all geological teaching subordinate to the literal interpretation of the early chapters of Genesis ; and in his *Reliquiae Diluvianae* (1823) he stoutly maintained the view that all the superficial deposits of the globe were the result of the Noachian deluge! He was indeed the great leader of the Catastrophists, and it is not surprising to find Lyell, while still under his influence, scoffing at 'the Huttonians[32].'

That Buckland greatly influenced Lyell in his youth, especially by inoculating him with his splendid enthusiasm for geology, there can be no doubt ; and Lyell, far as he departed in after life from the views of his teacher, never forgot his indebtedness to the Oxford professor. Even in 1832, in publishing the second edition of the first volume of his *Principles*, he dedicated it to Buckland, as one 'who first instructed me in the elements of geology, and by whose energy and talents the cultivation of science in the country has been so eminently promoted[33].'

On leaving Oxford in 1819, at the age of twenty-two, Lyell joined the Geological Society. What were the dominant opinions at that time on geological theory among the distinguished men, who were there

laying the foundations of stratigraphical geology, we
have already seen.  Lyell, in his frequent visits to the
continent, became a friend of the illustrious Cuvier,
whose strong bias for Catastrophism was so forcibly
shown in his writings and conversation.

What then, we may ask, were the causes which led
Lyell to abandon the views in which he had been
instructed, and to become the great champion of
Evolutionism?

It has often been assumed that Lyell was led by
the study of Hutton's works to adopt the 'Uniformitarian' doctrines.  But there is ample evidence that
such was not the case.  As late as the year 1839,
Lyell wrote of Hutton, 'Though I tried, I doubt
whether I fairly read half his writings, and skimmed
the rest[34]' ; and he emphatically assured Scrope 'Von
Hoff has assisted me most[35].'

The fact is certain that Lyell, quite independently,
arrived at the same conclusions as Hutton, *but by
totally different lines of reasoning.*

As early as 1817, when Lyell was only twenty
years of age, he visited the Norfolk coast and was
greatly impressed by the evidence of the waste of the
cliffs about Cromer, Aldborough, and Dunwich ; and
three years later we find him studying the opposite
kind of action of the sea in the formation of new
land at Dungeness and Romney Marsh.  All through
his life there may be seen the results of these early

studies in a tendency which he showed to *overrate marine action*; the chief defect in his early views consisting in not fully realising the importance of that subaerial denudation—of which Hutton was so great an exponent. But it was in his native county of Forfarshire that Lyell found the most complete antidote to the Catastrophic teachings. Buckland had taught him that the 'till' of the country had been thrown down, just 4170 years before, by the Noachian deluge : while Cuvier had asserted that the study of freshwater limestones proved them to differ from any recent deposit by their crystalline character, the absence of shells and the presence of plant-remains, as well as by the occasional occurrence in them of bands of flint. As the result of this, Cuvier and Brongniart had declared that *the freshwater of the ancient world possessed properties which are not observed in that of modern lakes* [36]. Lyell visited Kinnordy from time to time between 1817 and 1824, and found on his father's estate and other localities in Strathmore a number of small lakes, lying in hollows of the boulder clay. These were being drained and their deposits quarried for the purpose of 'marling' the land ; the excavations thus made showed that, under peat containing a boat hollowed out of the trunk of a tree, there were calcareous deposits, sometimes 16 to 20 feet in thickness, which passed into a rock, solid and crystalline

in character as the materials of the older geological
formations and containing the stems and fruits of the
freshwater plant *Chara* (Stone wort).

With the help of Robert Brown the botanist, and
of analyses made by Daubeny, with the advice of his
life-long friend, Faraday, Lyell was able to demon-
strate that from the waters of the Forfarshire lakes,
containing the most minute proportions of calcareous
salts, a limestone, identical in all respects with those
of the older rocks of the globe, had been deposited,
with excessive slowness, by the action of plant-life[37].
He was thus enabled to supply a complete refutation
of the views put forward by Buckland and Cuvier.

Thus while Hutton had been led to his conclusion
concerning evolution in the inorganic world, by
studying the waste going on in the weathered crags
and the flooded rivers of his native land, Lyell's
conversion to the same views was mainly brought
about by the study of changes due to the action of
the sea along the English coasts, and by studying the
evidence of constant, though slow, deposition of lime-
stone-rocks, by the seemingly most insignificant of
agencies.

Lyell however did not by any means neglect the
study of the action of rain and rivers. During his
visits to Forfarshire, he had his initials and the date
cut by a mason on many portions of the rocky river-
beds about his home. Fifty years afterwards (in

1874) I visited with him the several localities, to
ascertain what amount of waste had resulted from
the constant flow of water over these hard rocks.  It
was in most cases singularly small, the inscriptions
being still visible, though deprived of their sharpness;
even the sandy detritus carried along by the streams,
being buoyed up by the water, had not been able in
half a century to wear away a thickness of half-an-
inch of the hard rock.  The most singular result
we noticed was, that the leaden small shot fired by
sportsmen, in the Highland tracts, whence these
streams flowed, had collected in great numbers in
hollows formed by the young geologist's inscriptions.

By his father's request, Lyell after leaving Oxford
studied for the bar, but there is no doubt that his
main interest was in geological study.  He had made
the acquaintance of Dr Mantell, and carried on
a number of researches in the south of England
either alone or with that geologist[38].  Four years
after joining the Geological Society, in which he was
a constant worker, he became one of the secretaries.
This was in 1823 when he was only 26 years of age.
His frequent visits to Paris and to various parts of
the continent enabled him to exchange ideas with
many foreign naturalists, and it is clear from his
correspondence that at this early period he had
abandoned the Catastrophic doctrines of his teachers
and friends.

Let us now consider the outside influences which were at work on Lyell's mind in these early days. In the year 1818, the eminent palaeontologist Blumenbach induced the University of Göttingen to offer a prize for an essay on '*The investigation of the changes that have taken place in the earth's surface conformation since historic times, and the applications which can be made of such knowledge in investigating earth revolutions beyond the domain of history.*' A young German, Von Hoff, won the prize by a most able book, displaying great erudition, entitled *The History of those Natural Changes in the Earth's Surface, which are proved by Tradition.* The first volume of this work appeared in 1822, and treated of the results produced on the land by the action of the sea; the second volume, published in 1824, dealt with the effects of volcanoes and earthquakes. Von Hoff's learned work was confined to the collection of data from classical and other early authors bearing on these subjects, and to reasonings based on these records; for, unfortunately, he did not possess the means necessary for travelling and making observations in the districts described by him. Lyell acknowledges the great assistance afforded to him by these two volumes of Von Hoff's work, but, unlike that author, he was able to visit the various localities referred to, and to draw his own conclusions as to the nature of the changes which must have taken

place.  It is pleasant to be able to relate that the debt
which he owed to Von Hoff was fully repaid by Lyell;
for the learned German's third volume appeared after
the issue of the *Principles of Geology,* and as Zittel
assures us 'its influence on Von Hoff is quite apparent
in the third volume of his work [39].'

At this period, too, Lyell had the advantage of
travelling both on the continent and in various parts
of Great Britain with the eminent French geologist,
Constant Prevost, who had shown his courage by
opposing some of the catastrophic teachings of the
illustrious Cuvier himself.

Still more important to Lyell were the opportunities he enjoyed for comparing his conclusions
with those of Scrope, who had joined the Geological
Society in 1824, and became a joint secretary with
Lyell in the following year.  From both of them, in
their old age, I heard many statements concerning the
closeness and warmth of their friendship, and the
constant interchange of ideas which took place
between them at this time.

From Scrope, Lyell heard of the occurrence of
great beds of freshwater limestone in the Auvergne,
on a far grander scale than in Strathmore, with many
other facts concerning the geology of Central France,
which so greatly excited him as in the end to alter
all his plans concerning the publication of his own
book.  As soon as Scrope's great work on Auvergne

was published, Lyell undertook the preparation of
a review for the *Quarterly*—and this review was
a very able and discriminating production.

Although Lyell did not derive his views concerning terrestrial evolution directly from Hutton,
as is sometimes supposed, there were two respects
in which he greatly profited when he came to read
Hutton's work at a later date.

In the first place, he was very deeply impressed
by the necessity of avoiding the *odium theologicum*,
which had been so strongly, if unintentionally, aroused
by Hutton, of whom he wrote, 'I think he ran un-
necessarily counter to the feelings and prejudices of
the age. This is not courage or manliness in the
cause of Truth, nor does it promote progress. It
is an unfeeling disregard for the weakness of human
nature, for it is our nature (for what reason heaven
knows), but as *it is* constitutional in our minds, to
feel a morbid sensibility on matters of religious faith,
I conceive that the same right feeling which guards
us from outraging too violently the sentiments of our
neighbours in the ordinary concerns of the world and
its customs, should direct us still more so in this[40].'

In the second place, Lyell was warned by the fate
of Hutton's writings that it was hopeless to look
for success in combatting the prevailing geological
theories, unless he cultivated a literary style very
different from that of the *Theory of the Earth.*

4—2

Lyell's father had to a great extent guided his son's classical studies, and at Oxford, where Lyell took a good degree in classics, he practised diligently both prose and poetic composition. Lyell once told me that his tutor Dalby (afterwards a Dean) had put Gibbon's *Decline and Fall of the Roman Empire* into his hand with certain passages marked as 'not to be read.' When he had studied the whole work (of course including the marked passages) he said he conceived a profound admiration for the author's literary skill—and this feeling he retained throughout his after life. It is not improbable, indeed, that Lyell learned from Gibbon that a 'frontal attack' on a fortress of error is much less likely to succeed than one of 'sap and mine.' Lyell was always most careful in the composition of his works, sparing no pains to make his meaning clear, while he aimed at elegance of expression and logical sequence in the presentation of his ideas. The weakness of his eyes was a great difficulty to him, throughout his life, and, when not employing an amanuensis, he generally wrote stretched out on the floor or on a sofa, with his eyes close to the paper.

The relation of Lyell's views to those of Hutton, may best be described in the words of his contemporary, Whewell, whose remarks written immediately after the publication of the first volume of the *Principles*, lose nothing in effectiveness from the

evident, if gentle, note of sarcasm running through
them :—

'Hutton for the purpose of getting his continents above water,
or manufacturing a chain of Alps or Andes, did not disdain to call
in something more than common volcanic eruptions which we read
of in newspapers from time to time. He was content to have
a period of paroxysmal action—an extraordinary convulsion in
the bowels of the earth—an epoch of general destruction and
violence, to usher in one of restoration and life. Mr Lyell throws
away all such crutches, he walks alone in the path of his speculations ; he requires no paroxysms, no extraordinary periods ; he is
content to take burning mountains as he finds them ; and, with
the assistance of the stock of volcanoes and earthquakes now on
hand, he undertakes to transform the earth from any one of its
geological conditions to any other. He requires time, no doubt ;
he must not be hurried in his proceedings. But, if we will allow
him a free stage in the wide circuit of eternity, he will ask no
other favour ; he will fight his undaunted way through formations, transition and flötz—through oceanic and lacustrine
deposits ; and does not despair of carrying us triumphantly
from the dark and venerable schist of Skiddaw, to the alter-
nating tertiaries of the Isle of Wight, or even to the more recent
shell-beds of the Sicilian coasts, whose antiquity is but, as it were,
of yester-myriad of years[41].'

Never, surely, did words written in a tone of
banter constitute such real and effective praise !

But though it is certain that Lyell did not *derive*
his evolutionary views from Hutton, yet when he
came to write his historical introduction to the
*Principles,* he was greatly impressed by the proofs

of genius shown by the great Scotch philosopher, and equally by the brilliant exposition of those views by Playfair in his *Illustrations*. To the former he gave unstinted praise for the breadth and originality of his views, and to the latter for the eloquence of his writings—adopting quotations chosen from these last, indeed, as mottoes for his own work.

It is only just to add that for the violent prejudices excited by some of his contemporaries against Hutton's writings—as being directed against the theological tenets of the day and therefore subversive of religion—there is really no foundation whatever; and every candid reader of the *Theory of the Earth* must acquit its author of any such design. The passage quoted on page 51 could only have been written by Lyell at a time when he was still unacquainted with Hutton's works, and was misled by common report concerning them. It is interesting to note, however, that the passage occurs in a letter written in December 1827, that is after the first draft of the *Principles of Geology* had been 'delivered to the publisher,' and before the preparation of the historical introduction, which would appear to have led to the first perusal of Hutton's great work, and that of his brilliant illustrator, Playfair.

# CHAPTER VI

## 'THE PRINCIPLES OF GEOLOGY'

WE have seen that as early as the year 1817, when he visited East Anglia, Lyell began to experience vague doubts concerning the soundness of the 'Catastrophist' doctrines, which had been so strongly impressed upon him by Buckland. And these doubts in the mind of the undergraduate of twenty years of age gradually acquired strength and definiteness during his frequent geological excursions, at home and abroad, during the next ten years. At what particular date the design was formed of writing a book and attacking the predominant beliefs of his fellow-geologists, we have no means of ascertaining exactly; but from a letter written to his friend Dr Mantell, we find that at one time Lyell contemplated publishing a book in the form of 'Conversations in Geology[42],' without putting his name to it. This was probably suggested by the manner in which Copernicus and Galileo sought to circumvent theological opposition in the case of Astronomical Theory.

But this plan appears to have been soon abandoned; and by the end of the year 1827, when he had reached the age of thirty, Lyell had sent to the printer the first manuscript of the *Principles of Geology*, proposing that it should appear in the course of the following year in two octavo volumes[43].

A great and sudden interruption to this plan occurred however, for just at this time Lyell was engaged in writing his review for the *Quarterly* of Scrope's work on *The Geology of Central France*, and while doing this his interest was so strongly aroused by the accounts of the phenomena exhibited in the Auvergne, that he was led for a time to abandon the task of seeing his own book through the press; and, having induced Murchison and his wife to accompany him, set off on a visit to that wonderful district. He also felt that, before completing the second part of his book, he needed more information concerning the Tertiary formations, especially in Italy.

Lyell had been very early convinced of the supreme importance of travel to the geologist. In a letter to his friend Murchison he said :—'We must preach up travelling, as Demosthenes did "delivery" as the first, second and third requisites for a modern geologist, in the present adolescent state of the science[44].'

And Professor Bonney has estimated that so far did he himself practise what he preached, that no

less than one fourth of the period of his active life
was spent in travel[45].

The joint excursion of Lyell and Murchison to
the Auvergne was destined to have great influence
on the minds of these pioneers in geological research;
both became satisfied from their studies that, with
respect to the excavation of the valleys of the
country, Scrope's conclusions were irresistible; and
in a joint memoir this position was stoutly maintained by them.

It is interesting to notice the impression made by
these two great geologists on one another during this
joint expedition.

Murchison wrote that he had seen in Lyell 'the
most scrupulous and minute fidelity of observation
combined with close application in the closet and
ceaseless exertion in the field[46].'

But I recollect that Lyell once told me how
difficult Murchison found it to restrain himself from
impatience, when his companion's attention was
drawn aside by his entomological ardour. In an
early letter, indeed, we find that Murchison often
expressed a wish that Lyell's sisters had been with
them to attend to the insect-collecting and thus leave
Lyell free for geological work[47].

On the other hand, Lyell informed me that
Murchison had rendered him a great service in
showing how much a geologist could accomplish by

taking advantage of riding on horseback, and he
declared in his letters that he 'never had a better
man to work with than Murchison'; nevertheless he
ridiculed his 'keep-moving-go-it-if-it-kills-you' system
as—quoting from the elder Matthews—he called it[48].

On parting from Murchison and his wife, after the
Auvergne tour, Lyell proceeded to Italy and for more
than a year he was busy studying the Tertiary
deposits of Lombardy, the Roman states, Naples
and Sicily, and conferring with the Italian geologists
and conchologists. Thus it came about that he was
not free to resume the task of seeing the *Principles*
through the press till February 1829.

Immediately after his return to England Lyell
was compelled, with the assistance of his companion
Murchison, to defend their conclusions concerning
the excavations of valleys by rivers from a determined attack of Conybeare, who was backed up
by Buckland and Greenough; the old geologists
endeavoured to prove that the river Thames had
never had any part in the work of forming its
valley[49]. It is interesting to find that, on this
occasion, Sedgwick, who was in the chair, was so
far influenced by the arguments brought forward
by the young men, as to lend some aid to those who
had come to be called the 'Fluvialists,' in contradistinction to the 'Diluvialists'; he went so far as to
suggest that, with regard to the floods which the

Catastrophist invoked, it would be wiser at present to
'doubt and not dogmatise.'

To what extent the MS. of the *Principles*, sent
to the publisher in 1827, was added to and altered
two years later, we have no means of knowing ; but
that the work was to a great extent rewritten would
appear from a letter sent to Murchison by Lyell, just
before his return to England.   In it, he says :—

'My work is in part written, and all planned.   It
will not pretend to give even an abstract of all that
is known in geology, but it will endeavour to establish
*the principle of reasoning* in the science ; and all my
geology will come in as illustration of my views of
those principles, and as evidence strengthening the
system necessarily arising out of the admission of
such principles, which, as you know, are neither more
nor less than that *no causes whatever* have from the
earliest time to which we can look back to the present,
ever acted, but those that are *now acting*, and that
they never acted with different degrees of energy
from that which they now exert'; but in 1833, in
dedicating his third volume to Murchison, he refers
to the MS., completed in 1827, as a 'first sketch
only of my *Principles of Geology*[51].'

At one period, Lyell contemplated again delaying
publication till he had visited Iceland.   In the end,
however, after declining to act as professor of geology
in the new 'University of London'(University College),

he set himself down steadily to the task of seeing the
book through the press. It was at this time that
Lyell experienced a singular piece of good fortune,
comparable with that which befel Darwin thirty years
afterwards, by his book falling into the hands of a
very sympathetic reviewer. John Murray, who had
undertaken the publication of the *Principles*, was
also the publisher of the *Quarterly Review*, and
Lockhart, the editor of that publication, undertook
that an early notice of the book should appear, if the
proof-sheets were sent to the reviewer. Buckland
and Sedgwick were successively approached on the
subject of reviewing Lyell's book, but both declined
on the ground of 'want of time'; though I strongly
suspect that their real motive in refusing the task
was a disinclination to attack—as they would doubtless have felt themselves compelled to do—a valued
personal friend. Conybeare was, fortunately, thought
to be out of the question, as Lockhart said he
'promises and does not perform in the reviewing line.'

Very fortunately at this juncture, Lockhart, who
was in the habit of attending the Geological Society
and listening to the debates (for as he used to say
to his friends whom he took with him from the
Athenaeum, 'though I don't care for geology, yet I
*do* like to see the fellows fight') thought of Scrope.
Although he had practically retired from the active
work of the Geological Society at this time, Scrope

was known as an effective writer, and, happily for
the progress of science, he undertook the review of
Lyell's book.

Although, of course, Lyell had no voice in the
choice of a reviewer for the *Principles*, yet he could
not fail to rejoice in the fact that it had fallen to his
friend, who so strongly sympathised with his views,
to introduce it to the public. While the book was
being printed and the review of it was in preparation,
a number of letters passed between Lyell and Scrope,
and the latter, before his death, gave me the carefully
treasured epistles of his friend, with the drafts of
some of his replies. These letters, some of which
have been published, throw much light on the difficulties with which Lyell had to contend, and the
manner in which he strove to meet them.

As we have already seen, many of the leaders in
the Geological Society at that day besides being
strongly inclined to Wernerian and Cataclysmal views,
had an honest, however mistaken, dread lest geological research should lead to results, apparently
not in harmony with the accounts given in Genesis
of the Creation and the Flood. Lyell, as this correspondence shows, was most anxious to avoid exciting
either scientific or theological prejudice. He wrote,
'I conceived the idea five or six years ago' (that is
in 1824 or 5) that 'if ever the Mosaic geology could
be set down without giving offence, it would be in an

historical sketch[52],' and 'I was afraid to point the
moral...about Moses. Perhaps I should have been
tenderer about the Koran[53].' He further says 'full
*half* of my history and comments was cut out, and
even many facts, because either I, or Stokes, or
Broderip, felt that it was anticipating twenty or
thirty years of the march of honest feeling to declare
it undisguisedly[54].'

Under these circumstances the publication by
Scrope of his two long notices of the *Principles*
in the *Review* which was regarded as the champion
of orthodoxy, was most opportune. A very clear
sketch was given in these reviews of the leading
facts and the general line of argument; and at the
same time the allowing of prejudice or prepossession
to influence the judgment on such questions was very
gently deprecated[55].

But Scrope's reviews did not by any means
consist of an indiscriminate advocacy of Lyell's
views. In one respect—that of the great importance
of subaerial action as contrasted with marine action
—Scrope's views were at this time in advance of
those of Lyell, and he called especial attention to the
direct effects produced by rain in the earth-pillars
of Botzen. These Lyell had not at the time seen,
but took an early opportunity of visiting. Scrope,
too, was naturally much more speculative in his modes
of thought than Lyell, and argued for the probably

greater intensity in past times of the agencies causing geological change, and for the legitimacy of discussing the mode of origin of the earth. Lyell, like Hutton, argued that he saw '*no signs* of a beginning,' but his characteristic candour is shown when he wrote :—

'All I ask is, that at any given period of the past, don't stop enquiry, when puzzled, by a reference to a "beginning," which is all one with "another state of nature," as it appears to me. But there is no harm in your attacking me, provided you point out that it is the *proof* I deny, not the *probability* of a beginning[56].'

Lyell clearly foresaw the opposition with which his book would be met and wisely resolved not to be drawn into controversy. He wrote :—

'I daresay I shall not keep my resolution, but I will try to do it firmly, that when my book is attacked...I will not go to the expense of time in pamphleteering. I shall work steadily on Vol. II, and afterwards, if the work succeeds, at edition 2, and I have sworn to myself that I will not go to the expense of giving time to combat in controversy. It is interminable work[57].'

In order to maintain this resolve, Lyell, the moment the last sheet of the volume was corrected, set off for a four months' tour in France and Spain. While absent from England, he heard little of what was going on in the scientific world ; but, on his return, Lyell was told by Murray that in the three months before the *Quarterly Review* article appeared, 650 copies of the volume, out of the 1500 printed, had been sold, and he anticipated the disposal of many more, when the review came out. This expectation was realised and led to the issue of a second edition of the first volume, of larger size and in better type.

Lyell, from the first, had seen that it would be impossible to avoid the conclusion that the principles which he was advancing with respect to the inorganic world must be equally applicable to the organic world. At first he only designed to touch lightly on this subject, in the concluding chapters of his first volume, and to devote the second volume to the application of his principles to the interpretation of the geological record. He, however, found it impossible to include the chapters on changes in the organic world in the first volume and then decided to make them the opening portion of the second volume.

It is evident, however, that as the work progressed, the interest of the various questions bearing on the origin of species grew in his mind. While Lyell found it impossible to accept the explanation of origin suggested by Lamarck, he was greatly influenced by the arguments in favour of evolution advanced by that naturalist; and as he wrote chapter after chapter on the questions of the modification and variability of

species, on hybridity, on the modes of distribution of
plants and animals, and their consequent geographical
relations, and discussed the struggle of existence
going on everywhere in the organic world, in its
bearings on the question of 'centres of creation,' he
found the second volume growing altogether beyond
reasonable limits.  His intense interest in this part
of his work is shown by his remark, 'If I have succeeded so well with inanimate matter, surely I shall
make a lively thing when I have chiefly to talk of
living beings[58]?'

By December 1831, Lyell had come to the resolution to publish the chapters of his work which dealt
with the changes going on in the organic world as
a volume by itself.  This second volume of the
*Principles* he gracefully dedicated to his friend
Broderip, who had rendered him such valuable assistance in all questions connected with Natural History.

This volume appeared in January 1832, at the
same time that a second edition of the first volume
was also issued.  The reception of the second volume
by the public appears to have been not less favourable
than that of the first.

In March 1831, Lyell had accepted the Pro-
fessorship of Geology in King's College, London.
In addition to his desire to aid in the work of
scientific education, in which he had always taken so
great an interest, Lyell seems to have felt that the

task of presenting his views in a popular form would
be aided by his having to expound them to a miscel-
laneous audience. For two years, these lectures
were delivered, and attracted much attention; the
favourable impressions produced by them on a man of
the world have been recorded by Abraham Hayward,
and on more scientific thinkers by Harriet Martineau.

The third volume of the *Principles* was not
completed till a second edition of the second volume
had been issued. This third volume, appearing in
May 1833, dealt with the classification of the Tertiary
strata, to which Lyell had devoted so much labour,
studying conchology under Deshayes, and visiting all
the chief Tertiary deposits of Europe for the collection of materials. The application of the principles
enunciated in the two earlier volumes to the un-
ravelling of the past history of the globe, constituted
the chief task undertaken in this part of the great
work. But not a few controversial questions were
dealt with, and the famous 'metamorphic theory'
was advanced in opposition to the Wernerian hypothesis of 'primitive formations.' The volume was
appropriately dedicated to Murchison, who had been
Lyell's companion in the famous Auvergne excursion,
which had produced such an effect on his mind.

Within a twelvemonth, a third edition of the
whole work in four small volumes was issued, and in
the end no less than twelve editions of the *Principles*

*of Geology* were issued, in addition to portions
separately published under the titles of *Manual,
Elements,* and *Student's Elements of Geology,* of all
of which a number of editions have appeared. Lyell
was always the most painstaking and conscientious
of authors. He declared 'I must write what will be
read[59],' and he spared no labour in securing accuracy
of statement combined with elegance of diction. His
father, a good classical and Italian scholar, had done
much towards assisting him to attain literary excellence, and at Oxford, where he took a good degree
in classics, he was greatly impressed by the style of
Gibbon's writings, and practised both prose and
poetic compositions with great diligence.

Both Darwin and Huxley always maintained that
the real charm and power of Lyell's work are only to
be found in the *first edition*[60]. As new discoveries
were made or more effective illustrations of his views
presented themselves to his mind, passage after
passage in the work was modified by the author
or replaced by others; and the effects of these
constant changes—however necessary and desirable
in themselves—could not fail to be detrimental to
the book as a work of art. He who would form a
just idea of the greatness of Lyell's masterpiece,
must read the first edition, of course bearing in
mind, all the while, the state of science at the time
it was written.

# CHAPTER VII

## THE INFLUENCE OF LYELL'S WORKS

ALTHOUGH the *Principles of Geology* was received by the public with something like enthusiasm—due to the cogency of its reasoning and the charm of its literary style—there were not wanting critics who attacked the author on the ground of his heterodox views. It had come to be so generally understood, that every expression of geological opinion should, by way of apology, be accompanied by an attempt to 'harmonise' it with the early chapters of Genesis, that the absence of any references of this kind was asserted to be a proof of 'infidelity' on the part of the author.

But Lyell's sincere and earnest efforts to avoid exciting theological prejudice, and the striking illustrations, which he gave in his historical introduction, of the absurdities that had resulted from these prejudices in the past, were not without effect. This was shown in a somewhat remarkable manner in 1831, when, in response to an invitation given to him, he consented to become a candidate for the Chair of Geology at King's College, London, then recently founded.

The election was in the hands of an Archbishop, two Bishops and two Doctors of Divinity, and Lyell relates their decision, as communicated to him, in the following words :—

'They considered some of my doctrines startling enough, but could not find that they were come by otherwise than in a straight-forward manner, and (as *I* appeared to think) logically deducible from the facts, so that whether the facts were true or not, or my conclusions logical or otherwise, there was no reason to infer that I had made my theory from any hostile feeling towards revelation[61].'

The appointment was, in the end, made with only one dissentient, and it is pleasing to find that Cony-beare, the most determined opponent of Lyell's evolutionary views, was extremely active in his efforts in his support. The result was equally honourable to all parties, and affords a pleasing proof of the fact that in the half century which had elapsed since the persecution of Priestley and Hutton, theological rancour must have greatly declined. But while the reception of the *Principles of Geology* by the general public was of such a generally satisfactory character, Lyell had to acknowledge that his reasoning had but little effect in modifying the views of his distinguished contemporaries in the Geological Society.

The admiration felt for the author's industry and skill, in the collection and marshalling of facts and of the observations made by him in his repeated travels, were eloquently expressed by the generous Sedgwick, as follows :—

'Were I to tell "the author" of the instruction I received from every chapter of his work, and of the delight with which I rose from the perusal of the whole, I might seem to flatter rather than to speak the language of sober criticism ; but I should only give utterance to my honest sentiments. His work has already taken, and will long maintain a distinguished place in the philosophic literature of this country [62].'

Nevertheless, in the same address to the Geological Society, in which these words were spoken, Sedgwick goes on to argue forcibly against the doctrine of continuity, and to assert his firm belief in the occurrence of frequent interruptions of the geological record by great convulsions.

Whewell was equally enthusiastic with Sedgwick, concerning the value of the body of facts collected by Lyell, declaring that he had established a new branch of science, 'Geological Dynamics'; but he also believed with Sedgwick, that the evolutionary doctrine was as obnoxious to true science as he thought it was to Scripture.

These were the views of all the great leaders of

geological science at that day, and in 1834, after the
completion of the *Principles*, when a great discussion
took place in the Geological Society on the subject
of the effects ascribed by him to existing causes,
Lyell says that 'Buckland, De la Beche, Sedgwick,
Whewell, and some others treated them with as
much ridicule as was consistent with politeness in
my presence[63].'

It is interesting to be able to infer from Lyell's
accounts of these days, that the sagacious De la
Beche was beginning to weaken in his opposition to
evolutionary views, and that Fitton and John Phillips
were inclined to support him, but neither of them
was ready to come forward boldly as the champions
of unpopular opinions.  John Herschel, who sympathised with Lyell in all his opinions, was absent
at the Cape, Scrope was absorbed in the stormy
politics of that day, and it was not till Darwin
returned from his South American voyage in 1838,
that Lyell found any staunch supporter in the frequent lively debates at the Geological Society.

It is pleasing, however, to relate that this strong
opposition to his theoretical teachings, did not lessen
the esteem, or interfere with the friendship, felt for
Lyell by his contemporaries.  During all this time
he held the office of Foreign Secretary to the Society,
and in 1835 was elected President, retaining the office
for two years.

The general feeling of the old geologists with
respect to Lyell's opinions was very exactly expressed by Professor Henslow, when in parting from
young Darwin on his setting out on his voyage, he
referred to the recently published first volume of the
*Principles* in the following terms :—

'Take Lyell's new book with you and read it by
all means, for it is very interesting, but do not pay
any attention to it, except in regard to facts, for it is
altogether wild as far as theory goes.'

(I quote the words as repeated to me by Darwin,
in a conversation I had with him on August 7th, 1880,
of which I made a note at the time. Darwin has
himself referred to this conversation with Henslow
in his autobiography[64].)

Except in a few cases, this was the attitude
maintained by all the old geologists who were Lyell's
contemporaries. Even as late as 1895 we find the
amiable Prestwich protesting strongly against 'the
*Fetish* of uniformity[65],' and I well remember about
the same time being solemnly warned by a geologist
of the old school against 'poor old Lyell's fads.'

It was not, indeed, till a new generation of geologists had arisen, including Godwin-Austen, Edward
Forbes, Ramsay, Jukes, Darwin, Hooker and Huxley,
that the real value and importance of Lyell's teaching
came to be recognised and acknowledged.

The most important influence of Lyell's great

work is seen, however, in the undoubted fact that it inspired the men, who became the leaders in the revolution of thought which took place a quarter of a century later in respect to the organic world. Were I to assert that if the *Principles of Geology* had not been written, we should never have had the *Origin of Species*, I think I should not be going too far : at all events, I can safely assert, from several conversations I had with Darwin, that he would have most unhesitatingly agreed in that opinion.

Darwin's devotion to his 'dear master' as he used to call Lyell, was of the most touching character, and it was prominently manifested in all his geological conversations. In his books and in his letters he never failed to express his deep indebtedness to his 'own true love' as he called the *Principles of Geology*. In what was Darwin's own most favourite work, the *Narrative of the Voyage of the Beagle*, he wrote 'To Charles Lyell, Esq., F.R.S., this second edition is dedicated with grateful pleasure, as an acknowledgment that the chief part of whatever scientific merit this Journal and the other works of the author may possess, has been derived from studying the well-known, admirable *Principles of Geology*.'

How Lyell's first volume inspired Darwin with his passion for geological research, and how his second volume was one of the determining causes in turning his mind in the direction of Evolution, we shall

see in the sequel.  In 1844, Darwin wrote to Leonard
Horner how 'forcibly impressed I am with the
infinite superiority of the Lyellian School of Geology
over the continental,' he even says, 'I always feel as
if my books came half out of Lyell's brain'; adding
'I have always thought that the great merit of the
*Principles* was that it altered the whole tone of one's
mind, and therefore that, when seeing a thing never
seen by Lyell one yet saw it partially through his
eyes[66].'  About the same time Darwin wrote, 'I am
much pleased to hear of the call for a new edition of
the *Principles*: what glorious good that work has
done[67]!'  And in the *Origin of Species* he gives his
deliberate verdict on the book, referring to it as
'Lyell's grand work on the *Principles of Geology,*
which the future historian will recognise as having
produced a revolution in Natural Science[68].'

Darwin seemed always afraid, such was his
sensitive and generous nature, that he did not
sufficiently acknowledge his indebtedness to Lyell.
He wrote to his friend in 1845 :

'I have long wished not so much for your sake as for my own
feelings of honesty, to acknowledge more plainly than by mere
reference, how much I geologically owe you.  Those authors,
however, who like you educate people's minds as well as teach
them special facts, can never, I should think, have full justice
done them except by posterity, for the mind thus insensibly
improved can hardly perceive its own upward ascent.'

Very heartily, as I can bear witness from long intercourse with him, was this deep affection of Darwin reciprocated by the man who was addressed by him in his letters as 'Your affectionate pupil.' But a stranger who conversed with Lyell would have thought that he was the junior and a disciple ; so profound was his reverence for the genius of Darwin.

There can be no doubt that Lyell's extreme caution in statement, and his candour in admitting and replying to objections, had much to do with his acquirement of that authority with general, no less than with scientific, readers, which he so long enjoyed. In his candour he resembled his friend Darwin ; but his caution was carried so far that, even after full conviction had entered his mind on a subject, he would still hesitate to avow that conviction.  He was always obsessed by a feeling that there still *might be* objections, which he had not foreseen and met, and therefore felt it unsafe to declare himself.  No doubt the peculiarly trying circumstances under which his work was written—a seemingly hopeless protest against ideas held unswervingly by teachers and fellow-workers—led to the creation in him of this habit of mind.

Darwin, with all his candour, was of a far more sanguine and optimistic temperament than Lyell, and the difference between them, in this respect, often comes out in their correspondence.

Thus Darwin, from the horrors he had witnessed in South America, had come to entertain a most fanatical hatred of slavery—his abhorrence of which he used to express in most unmeasured terms.  Lyell, in his travels in the Southern United States, was equally convinced of the undesirability of the institution ; but he thought it just to state the grounds on which it was defended, by those who had been his hosts in the Slave-states.  Even this, however, was too much for Darwin, and he felt that he must 'explode' to his friend 'How could you relate so placidly that atrocious sentiment' (it was of course only quoted by Lyell) 'about separating children from their parents ; and in the next page speak of being distressed at the whites not having prospered :  I assure you the contrast made me exclaim out.  But I have broken my intention (that is not to write about the matter), so no more of this odious deadly subject[69].'

It was just the same in their mode of viewing scientific questions.  Thus in 1838, while they were in the midst of the fierce battle with the 'Old Guard' at the Geological Society, Lyell wrote to his brother-in-arms as follows :—

'I really find, when bringing up my Preliminary Essays in *Principles* to the science of the present day, so far as I know it, that the great outline, and even most of the details, stand so uninjured, and in many cases they are so much strengthened by

new discoveries, especially by yours, that we may begin to hope
that the great principles there insisted on will stand the test of
new discoveries[70].'

To which the younger and more ardent Darwin
warmly replied :—

'*Begin to hope*: why, the *possibility* of a doubt has never
crossed my mind for many a day. This may be very unphilo-
sophical, but my geological salvation is staked on it......it makes
me quite indignant that you should talk of *hoping*[71].'

When talking with Lyell at this time about the
opposition of the old school of geologists to their
joint views, Darwin said, 'What a good thing it
would be if every scientific man was to die at sixty
years old, as afterwards he would be sure to oppose
all new doctrines[72].'

In conversations that I had with him late in life,
Darwin several times remarked to me, that 'he had
seen so many of his friends make fools of themselves
by putting forward new theoretical views in their old
age, that he had resolved quite early in life, never to
publish any speculative opinions after he was sixty.'
But both in conversation and in his writings he always
maintained that Lyell was an exception to all such
rules, seeing that at last he adopted the theory of
Natural Selection in his old age, thus displaying the
most 'remarkable candour.'

All who had the pleasure of discussing geological

questions with Lyell will recognise the truth of the portrait drawn of his old friend by Darwin, about a year before his own death.

He says :—

'His mind was characterised, as it appeared to me, by clearness, caution, sound judgment, and a good deal of originality. When I made a remark to him on Geology, he never rested until he saw the whole case clearly, and often made me see it more clearly than I had done before.'

And he sums up his admiration of the 'dear old master' in the words

'The science of Geology is enormously indebted to Lyell— more so, as I believe, than to any other man who ever lived[73].'

Alfred Russel Wallace is scarcely less emphatic than Charles Darwin himself in his expression of affection and admiration for Lyell, and his indebtedness to the *Principles of Geology*.

In his Autobiography, Wallace writes :—
'With Sir Charles I soon felt at home, owing to his refined and gentle manners, his fund of quiet humour, and his intense love and extensive knowledge of natural science. His great liberality of thought and wide general interests were also attractive to me ; and although when he had once arrived at a definite conclusion, he held by it very tenaciously until a considerable body of well-ascertained facts could be adduced against it, yet he was always willing to listen to the arguments of his opponents, and to give them careful and repeated consideration[74].'

Of the influence of the *Principles of Geology* in leading him to evolution, he wrote :

'Along with Malthus I had read, and been even more deeply impressed by, Sir Charles Lyell's immortal *Principles of Geology* ; which had taught me that the inorganic world—the whole surface of the earth, its seas and lands, its mountains and valleys, its rivers and lakes, and every detail of its climatic conditions—were and always had been in a continual state of slow modification. Hence it became obvious that the forms of life must have become continually adjusted to these changed conditions in order to survive.  The succession of fossil remains throughout the whole geological series of rocks is the record of the change ; and it became easy to see that the extreme slowness of these changes was such as to allow ample opportunity for the continuous automatic adjustment of the organic to the inorganic world, as well as of each organism to every other organism in the same area, by the simple processes of "variation and survival of the fittest."  Thus was the fundamental idea of the "origin of species" logically formulated from the consideration of a series of well ascertained facts[76].'

Nor were the two men (who, like Aaron and Hur so steadily sustained the hands of Darwin in his long vigil), behind the two authors of Natural Selection themselves in their devotion to Lyell.  How touching is Hooker's tribute of affection on the death of his friend, ' My loved, my best friend, for well nigh forty years of my life.  To me the blank is fearful, for it never will, never can be filled up.  The most generous sharer of my own and my family's hopes, joys, and

sorrows, whose affection for me was truly that of a
father and brother combined[76]."

And Huxley speaking of Lyell, the day after his
death said, 'Sir Charles Lyell would be known in
history as the greatest geologist of his time. Some
days ago I went to my venerable friend, and put
before him the results of the *Challenger* expedition.
Nothing could then have been more touching than
the conflict between the mind and the material body,
the brain clear and comprehending all; while the
lips could hardly express the views which the busy
mind formed[77].'

How well do I recollect my last visit to Lyell a
day or two after this farewell interview with Huxley,
the glow of gratitude which lighted up the noble
features as with trembling lips he told me how
'Huxley had repeated his whole Royal Institution
lecture at his bedside.'

Huxley was a most devoted student of Lyell.
Speaking to his fellow geologists in 1869 he said,
'Which of us has not thumbed every page of the
*Principles of Geology*[78]?' and writing in 1887 on the
reception of the *Origin of Species*, he said :—

'I have recently read afresh the first edition of the *Principles
of Geology*; and when I consider that this remarkable book had
been nearly thirty years in everybody's hands, and that it brings
home to any reader of ordinary intelligence a great principle and a
great fact—the principle, that the past must be explained by the

present, unless good cause be shown to the contrary; and the fact, that, so far as our knowledge of the past history of life on our globe goes, no such cause can be shown—I cannot but believe that Lyell, for others, as for myself, was the chief agent in smoothing the road for Darwin.  For consistent uniformitarianism postulates evolution as much in the organic as in the inorganic world.  The origin of a new species by other than ordinary agencies would be a vastly greater 'catastrophe' than any of those which Lyell successfully eliminated from sober geological speculation[70].'

How strongly Lyell had become convinced, as early as 1832, of the truth and importance of the doctrine of Evolution—in the *organic* as well as in the inorganic world—in spite of his emphatic rejection of the theory of Lamarck, we shall show in the next chapter.  It was this conviction, as we shall see, which led to his friendly encouragement of Darwin in his persevering investigations and to his constant solicitude that the results of his friend's labours should not be lost through delay in their publication.

# CHAPTER VIII

## EARLY ATTEMPTS TO ESTABLISH THE DOCTRINE
## OF EVOLUTION FOR THE ORGANIC WORLD

In studying the history of Evolutionary ideas, it is necessary to keep in mind that there are two perfectly distinct lines of thought, the origin and development of which have to be considered.

*First.* The conviction that species are not immutable, but that, by some means or other, new forms of life are derived from pre-existing ones.

*Secondly.* The conception of some process or processes, by which this change of old forms into new ones may be explained.

Buffon, Kant, Goethe, and many other philosophic thinkers, have been more or less firmly persuaded of the truth of the first of these propositions ; and even Linnaeus himself was ready to make admissions in this direction. It was impossible for anyone who was convinced of the truth of the doctrine of continuity or evolution in the *inorganic* world, to avoid the speculation that the same arguments by which the truth of that doctrine was maintained must apply also to the *organic* world.

Hence we find that directly the *Principles of Geology* was published, thinkers, like Sedgwick and Whewell, at once taxed Lyell with holding that ' the creation of new species is going on at the present day,' and Lyell replied to the latter :—

'It was impossible, I think, for anyone to read my work and not to perceive that my notion of uniformity in the existing causes of change always implied that they must for ever produce an endless variety of effects, *both in the animate and inanimate world*[80].'

And to Sedgwick, Lyell wrote :—

'Now touching my opinion,' concerning the creation of new species at the present day, 'I have no right to object, *as I really entertain it*, to your controverting it ; at the same time you will see, on reading my chapter on the subject, that I have studiously avoided laying down the doctrine dogmatically as capable of proof. I have admitted that we have only data for *extinction*, and I have left it to be inferred, instead of enunciating it even as my opinion, that the place of lost species is filled up (as it was of old) from time to time by new species. I have only ventured to say that had new mammalia come in, we could hardly have hoped to verify the fact[81].'

That Lyell was convinced of the truth of the doctrine of the evolution of species is shown by his correspondence with friends and sympathisers like Scrope and John Herschel. But he wrote:

'If I had stated....the possibility of the introduction or origination of fresh species being a natural, in contradistinction to a miraculous process, I should have raised a host of prejudices against me, which are unfortunately opposed at every step to any philosopher who attempts to address the public on these mysterious subjects[82].'

That Lyell was justified in not increasing the difficulties which would retard the reception of his views, by introducing matter, which he still regarded as of a more or less speculative character, I think everyone will be prepared to admit. Darwin had to contend with the same difficulty in writing the *Origin of Species*. To have included the question of the origin of mankind *prominently* in that work would have raised an almost insurmountable barrier to its reception. He says in his autobiography, 'I thought it best, in order that no honourable man should accuse me of concealing my views, to add that by the work "light would be thrown on the origin of man and his history." It would have been useless and injurious to the success of the book to have paraded, without giving evidence, my conviction with respect to his origin[83].'

Huxley and Haeckel have both borne testimony to the fact that Lyell, at the time he wrote the *Principles*, was firmly convinced that new species had originated by evolution from old ones. Indeed in a letter to John Herschel in 1836 he goes very far

in the direction of anticipating the lines in which
enquiries on the *method* of evolution must proceed,
having even a prevision of the doctrine of *mimicry*,
long afterwards established by Bates and others.
Lyell wrote :—

'In regard to the origination of new species, I am very glad
to find that you think it probable that it may be carried on
through the intervention of intermediate causes. I left this rather
to be inferred, not thinking it worth while to offend a certain class
of persons by embodying in words what would only be a speculation......One can in imagination summon before us a small part
at least of the circumstances that must be contemplated and
foreknown, before it can be decided what powers and qualities a
new species must have in order to enable it to endure for a given
time, and to play its part in due relation to all other beings
destined to coexist with it, before it dies out......It may be seen
that unless some slight additional precaution be taken, the species
about to be born would at a certain era be reduced to too low a
number. There may be a thousand modes of ensuring its
duration beyond that time ; one, for example, may be the
rendering it more prolific, but this would perhaps make it press
too hard upon other species at other times. Now if it be an
insect it may be made in one of its transformations to resemble a
dead stick, or a leaf, or a lichen, or a stone, so as to be somewhat
less easily found by its enemies ; or if this would make it too
strong, an occasional variety of the species may have this
advantage conferred on it ; or if this would be still too much,
one sex of a certain variety. Probably there is scarcely a dash
of colour on the wing or body of which the choice would be quite
arbitrary, or which might not affect its duration for thousands of
years. I have been told that the leaf-like expansions of the
abdomen and thighs of a certain Brazilian Mantis turn from

green to yellow as autumn advances, together with the leaves of plants among which it seeks its prey. Now if species come in succession, such contrivances must sometimes be made, and such relations predetermined between species, as the Mantis, for example, and plants not then existing, but which it was foreseen would exist together with some particular climate at a given time. But I cannot do justice to this train of speculation in a letter, and will only say that it seems to me to offer a more beautiful subject for reasoning and reflecting on, than the notion of great batches of new species all coming in and afterwards going out at once[84].'

We have cited this very remarkable passage, as it affords striking evidence of how deeply Lyell had thought on this great question at a very early period. Nevertheless it is certain that when he wrote the second volume of the *Principles,* he had not been able to satisfy himself that any hypothesis of the *mode* of evolution, that had up to that time been suggested, could be regarded as satisfactory.

The only serious attempt to *explain* the derivation of new species from old ones that came before Lyell was that of the illustrious Lamarck.

Very noteworthy was the work of that old wounded French soldier, afflicted in his later years as he was by blindness. By his early labours, Lamarck had attained a considerable reputation as a botanist, and later in life he turned his attention to zoology, and then to palaeontology and geology. In zoology, he did for the study of invertebrate

animals what his great contemporary Cuvier was
accomplishing for the vertebrates ; but, with regard
to the origin of species, he arrived at conclusions
directly at variance with those of his distinguished
rival.

We are indebted to Professor Osborn[85] for calling
attention to that remarkable, but little known work
of Lamarck's—*Hydrogéologie*—published in 1802,
seven years before his *Philosophie Zoologique* appeared. This work is especially interesting as showing
to how great an extent—as in the case of Darwin,
Wallace and others—it was geological phenomena
which played an important part in leading Lamarck
to evolutionary convictions. "In Geology," Professor
Osborn writes,

'Lamarck was an ardent advocate of uniformity, as against
the Cataclysmal School. The main principles are laid down in
his *Hydrogéologie*, that all the revolutions of the earth are extremely slow. "For Nature," he says, "time is nothing. It is never
a difficulty, she always has it at her disposal; and it is for her
the means by which she has accomplished the greatest as well as
the least results[86]."'

On the subject of subaerial denudation (the action
of rain and rivers in wearing down the earth's surface),
Lamarck's views were as clear and definite as those
of Hutton himself; though it is almost certain that
he could never have seen, or even heard of, the
writings of the great Scottish philosopher. On some

other questions of geological dynamics, however, it must be confessed that Lamarck's views and speculations were rather crude and unsatisfactory.

In his *Philosophie Zoologique*, published in the same year that Charles Darwin was born (1809), Lamarck brought forward a great body of evidence in favour of evolution, derived from his extensive knowledge of botany, zoology and geology. He showed how complete was the gradation between many forms ranked as species, and how difficult it was to say what forms should be classed as 'varieties' and what as 'species.'

But when he came to indicate a possible method by which one species might be derived from another, he was less happy in his suggestions. He recognised the value of the evidence derived from the study of the races which have arisen among domestic animals, and from the crossing of different forms. But his main argument was derived from the acknowledged fact that use or disuse may cause the development or the partial atrophy of organs—the case of the 'blacksmith's arm.' Unfortunately some of the suggestions made by Lamarck, in this connexion— like that of the elongation of the giraffe's neck to enable it to browse on high trees—were of a kind that made them very susceptible to ridicule. His theory was of course dependent on the admission that acquired characters were transmitted from parents to

children, and in the absence of any suggestion of
'selection,' it did not appeal strongly to thinkers on
this question.

Lyell first became acquainted with the writings
of Lamarck in 1827. As he was returning from the
Oxford circuit for the last time—having now resolved
to give up law and devote himself to geological work
exclusively—he wrote to his friend Mantell as
follows:—

'I devoured Lamarck *en voyage*......his theories delighted me
more than any novel I ever read, and much in the same way, for
they address themselves to the imagination, at least of geologists
who know the mighty inferences which would be deducible were
they established by observations. But though I admire even his
flights, and feel none of the *odium theologicum* which some
modern writers in this country have visited him with, I confess I
read him rather as I hear an advocate on the wrong side, to know
what can be made of the case in good hands. I am glad he has
been courageous enough and logical enough to admit that his
argument, if pushed as far as it must go, if worth anything, would
prove that men may have come from the Ourang-Outang. But
after all, what changes species may really undergo! How
impossible will it be to distinguish and lay down a line, beyond
which some of the so-called extinct species have never passed
into recent ones. That the earth is quite as old as he supposes,
has long been my creed, and I will try before six months are over
to convert the readers of the *Quarterly* to that heterodox
opinion[87].'

Lyell was at that time at work on his review for
the *Quarterly* of Scrope's *Central France,* and was

also completing the 'first sketch' of the *Principles*.
But it is evident that as the result of continued study
of Lamarck's book, Lyell found it, in spite of its
fascination, to embody a theory which he could not
but regard as unsound and not calculated to prove a
solution of the great mystery of evolution.   Accordingly when the second volume of the *Principles* was
issued in 1832, it was found to contain in its opening
chapters a very trenchant criticism of Lamarck's
theory.

It is only fair to remember, however, that in
1863, after Lyell had accepted the theory of Natural
Selection he wrote to Darwin :

'When I came to the conclusion that after all Lamarck was
going to be shown to be right, and that we must "go the whole
orang" I re-read his book, and remembering when it was written, I
felt I had done him injustice[88].'

It is interesting also to notice that Darwin, like
Lyell, gradually came to entertain a higher opinion
of the merit of Lamarck's works, than he did on his
first perusal of them.   In 1844, Darwin wrote to
Hooker, 'Heaven forfend me from Lamarck non-
sense !' and in the same year he speaks of Lamarck's
book as 'veritable rubbish,' an 'absurd though
clever work[89].'   When, after the publication of the
*Origin of Species*, Lyell referred to the *conclusions*
arrived at in that work as similar to those of

Lamarck, Darwin expressed something like indignation, and he wrote to their 'mutual friend'
Hooker, 'I have grumbled a bit in my answer to
him' (Lyell) 'at his *always* classing my book as a
modification of Lamarck's, which it is no more than
any author who did not believe in the immutability
of species[90].' In this case, as is so frequently seen in
the writings of Darwin, it is evident that he attaches
infinitely less importance to the establishment of the
*fact* of the evolution of species, than to the demonstration of a possible *mode of origin* of that evolution.
But that later in life Darwin came to take a more
indulgent view of the result of Lamarck's labours is
shown by a passage in his 'Historical Sketch'
prefixed to the *Origin*, in 1866. Lamarck, he says,
'first did the eminent service of arousing attention
to the probability of all change in the organic world,
as well as in the inorganic world, being the result of
law and not of miraculous interposition[91].'

In the opinion of Dr Schwalbe and others there
are indications in Darwin's later writings that he had
come into much closer relation with the views of
Lamarck, than was the case when he wrote the
*Origin*[92].

It is interesting, however, to note that Erasmus
Darwin, the grandfather of Charles, published
independently and contemporaneously, views on the
nature and causes of evolution in striking agreement

with those of Lamarck; but perhaps the poetical
form, in which he chose to embody his ideas, led to
their receiving less attention than they deserved.

As is now well known a number of writers during
the earlier years of the nineteenth century published
statements in favour of evolutionary views, and in
several cases the theory of natural selection was
more or less distinctly outlined.   In addition to
Geoffroy and Isidore Saint Hilaire and d'Omalius
d'Halloy on the continent, a number of writers
in this country, such as Dr Wells, Mr Patrick
Matthew, Dr Pritchard, Professor Grant, Dean
Herbert, all expressed views in favour of evolution,
even, in some cases, foreshadowing Natural Selection
as the method.   But these authors attached so little
importance to their suggestions, that they did not
even take the trouble to place them on permanent
record, and it is certain that neither Lyell nor
Darwin was acquainted with their writings at the
time they were themselves working at the subject.

There was indeed one work which, during the
time that the *Origin of Species* was in preparation,
attracted much popular attention.   In 1844, Robert
Chambers, who was favourably known as the author
of some geological papers, wrote a book which
excited a great amount of attention—the well-known
*Vestiges of Creation.*   This work was a very bold
pronouncement of evolutionary views.   Beginning

with a statement of the nebular hypothesis of Kant
and Laplace, it discussed the question of the origin
of life—when life became possible on a cooling
globe—and, arguing strongly in favour of the view
that all plants and animals, as the conditions under
which they existed change, had given rise to new
forms, better adapted to their environment, insisted
that the whole living creation had been gradually
developed from the simplest types.

Chambers published his book anonymously, being
naturally afraid of the prejudices that would be
excited against him—especially in his own country—
by a work so outspoken, and it was not till after his
death that its authorship was definitely known.

The *Vestiges of Creation* met with very different
receptions at the hands of the general public and
from the scientific world, at the time it was published.
The former were startled but captivated by its fear-
less statements and suggestive lines of thought;
while the latter were repelled and incensed by the
want of judgment, too frequently shown, in accept-
ing as indisputable, facts and experiments which
really rested on a very slender basis or none at all.
So popular was the book, however, that it passed
through twelve editions, the last being published
after the appearance of the *Origin of Species*.

It is interesting to read Darwin's judgment in
later life on this once famous book; he says :—

'The work from its powerful and brilliant style, though displaying in the earlier editions little accurate knowledge and a great want of scientific caution, immediately had a very wide circulation.  In my opinion it has done excellent service in this country in calling attention to the subject, in removing prejudice, and in thus preparing the ground for the reception of analogous views[93].'

If we enquire what was the attitude of scientific naturalists towards the doctrine of Evolution, immediately before the occurrence of the events to be recorded in the next chapter, we shall find some diversity of opinion to exist.  The late Professor Newton, an eniment ornithologist, has asserted that, at this period, many systematic zoologists and botanists had begun to feel great 'searchings of heart' as to the possibility of maintaining what were the generally prevalent views concerning the reality and immutability of species.  Huxley, however, declared that he and many contemporary biologists were ready to say 'to Mosaists and Evolutionists a plague to both your houses!' and were disposed to turn aside from an interminable and fruitless discussion, to labour in the fields of ascertainable fact[94].

# CHAPTER IX

## DARWIN AND WALLACE: THE THEORY OF
## NATURAL SELECTION

CHARLES DARWIN was the grandson of Erasmus
Darwin, who, as we have seen, arrived independently
at conclusions concerning the origin of species very
similar to those of Lamarck, and embodied his views
in poems, which, at the time of their publication,
achieved a considerable popularity. In the younger
philosopher, however, imagination was always kept in
subjection by a determination to '*prove* all things'
and 'to hold fast that which is good'; though, in
other respects, there were not wanting indications
of the existence of hereditary characteristics in the
grandson.

Born at Shrewsbury and educated in the public
school of that town, Charles Darwin from the first
exhibited signs of individuality in his ideas and his
tastes. The rigid classical teaching of his school did
not touch him, but, with the aid of his elder brother,
he surreptitiously started a chemical laboratory in a

garden tool-house.   From his earliest infancy he was
a collector, first of trifles, like seals and franks, but
later of stones, minerals and beetles.

At the outset, only the desire to possess new
things animated him, then a wish to put names to
them, but, at a very early period, a passion arose for
learning all he could about them.   Thus when only
9 or 10 years of age, he had 'a desire of being able
to know something about every pebble in front of
the hall-door,' and at 13 or 14, when he heard the
remark of a local naturalist, 'that the world would
come to an end before anyone would be able to
explain how' a boulder (the 'bell-stone' of local-fame)
came to be brought from distant hills—the lad had such
a deep impression made on his mind, that he says in
after life, 'I *meditated* over this wonderful stone[95].'

At the age of 16, he was sent to Edinburgh
University to prepare himself for the work of a
doctor—the profession of his father and grandfather.
But here his independence of character again asserted
itself.   He found most of the lectures 'intolerably
dull,' so he occupied himself with other pursuits,
making many friendships among the younger
naturalists and doing a little in the way of biological
research himself.

That he was not altogether destitute of ambition
in the eyes of his companions, however, is, I think,
indicated by an amusing circumstance.   In the

library of Charles Darwin, which is carefully preserved at Cambridge, there is a copy of Jameson's
*Manual of Mineralogy*, published in 1821, which
was evidently used by the young student in his class-
work at Edinburgh.   In this a quizzical fellow-student
has written 'Charles Darwin Esq., M.D., F.R.S.'—
mischievously adding 'A.S.S.'!   Even for geology,
the science to which in all his after life he became so
deeply devoted, young Darwin conceived the most
violent aversion; and as he listened to Jameson's
Wernerian outpourings at Salisbury Crags, he
'determined never to attend to geology,' registering
the terrible vow 'never as long as I lived to read a
book on Geology, or in any way to study the science[96].'

As it became evident that Charles Darwin would
never make a doctor, his father, after two years' trial,
sent him to Cambridge with the object of his
qualifying for a clergyman.   But at Christ's College,
in that University, he again took his own line—which
was not that of divinity—riding, shooting and beetle-
hunting being his chief delights.   Nevertheless, at
Cambridge as at Edinburgh, he seems to have shown
an appreciation for good and instructive society, and
in Henslow, the judicious and amiable Professor of
Botany, the young fellow found such sympathy and
kindly help that he came to be distinguished as 'the
man who walks with Henslow[97].'

After achieving a 'pass degree,' Darwin went

back to the University for an extra term, and by the
advice of Henslow began to 'think about' the
despised Science of Geology.  He was introduced to
that inspiring teacher, Sedgwick, with whom he
made a geological excursion into Wales ; but though
he said he 'worked like a tiger' at geology, yet he,
when he got the chance of shooting on his uncle's
estate, had to make the confession, 'I should have
thought myself mad to give up the first days of
partridge-shooting for geology or any other science[98].'

There is a sentence in one of the letters written
at this time which suggests that, even at this early
period in his geological career, Darwin had begun to
experience some misgivings concerning the catastrophic doctrines of his teachers and contemporaries.
He says :—

'As yet I have only indulged in hypotheses, but they are
such powerful ones that I suppose, if they were put into action
but for one day, the world would come to an end[99].'

Was he not poking fun at other hypotheses
besides his own ?

Darwin's real scientific education began when,
after some hesitation on his father's part, he was
allowed to accept the invitation, made to him through
his friend Henslow, to accompany, at his own expense,
the surveying ship *Beagle* in a cruise to South
America and afterwards round the world.  In the

narrow quarters of the little 'ten-gun brig,' he
learned methodical habits and how best to economise
space and time; during his long expeditions on
shore, rendered possible by the work of a surveying
vessel, he had ample opportunities for observing and
collecting; and, above all, the absence of the
distractions from quiet meditation, afforded by a
long sea-voyage, proved in his case invaluable.
Very diligently did he work, accumulating a vast
mass of notes, with catalogues of the specimens he
sent home from time to time to Henslow. He had
received no careful biological training, and Huxley
considered that the voluminous notes he made on
zoological subjects were almost useless[100]. Very
different was the case, however, with his geological
notes. He had learned to use the blowpipe, and
simple microscope, as well as his hammer and
clinometer; and the notes which he made concerning
his specimens, before packing them up for Cambridge,
were at the same time full, accurate and suggestive.

  Darwin has recorded in his autobiography the
wonderful effect produced on his mind by the reading
of the first volume of Lyell's *Principles*—an effect
very different from that anticipated by Henslow[101].
From that moment he became the most enthusiastic
of geologists, and never fails in his letters to insist on
his preference for geology over all other branches of
science. Again and again we find him recording

observations that he thinks will 'interest Mr Lyell'
and he says in another letter :—

'I am become a zealous disciple of Mr Lyell's views, as
known in his admirable book.  Geologising in South America,
I am tempted to carry parts to a greater extent even than he
does[102].'

Before reaching home after his voyage, the
duration of which was fortunately extended from two
to five years, he had sent home letters asking to be
elected a fellow of the Geological Society ; and,
immediately on his arrival, he gave up his zoological
specimens to others and devoted his main energies
for ten years to the working up of his geological
notes and specimens.

It may seem strange that the grandson of Erasmus
Darwin should in early life have felt little or no
interest in the question of the 'Origin of Species,' but
such was certainly the case.  He tells us in his
autobiography that he had read his grandfather's
*Zoonomia* in his youth, without its producing any
effect on him, and when at Edinburgh he says he
heard his friend Robert Grant (afterwards Professor
of Zoology in University College, London) as they
were walking together 'burst forth in high admiration of Lamarck and his views on Evolution'—yet
Darwin adds 'I listened in silent astonishment, and
as far as I can judge without any effect on my
mind[103].'

The reason of this indifference towards his grandfather's works is obvious. All through his life, Darwin, like Lyell, showed a positive distaste for all speculation or theorising that was not based on a good foundation of facts or observations. In this respect, the attitude of Darwin's mind was the very opposite of that of Herbert Spencer—who, Huxley jokingly said, would regard as a 'tragedy'—'the killing of a beautiful theory by an ugly fact.' Darwin tells us himself that, while on his first reading of *Zoonomia* he 'greatly admired' it— evidently on literary grounds—yet 'on reading it a second time after an interval of ten or fifteen years, I was much disappointed ; *the proportion of speculation being so large to the facts given.*' Huxley who knew Charles Darwin so well in later years said of him that :—

'He abhors mere speculation as nature abhors a vacuum. He is as greedy of cases and precedents as any constitutional lawyer, and all the principles he lays down are capable of being brought to the test of observation and experiment[104].'

What then, we may ask, were the facts and observations which turned Darwin's mind towards the great problem that came to be the work of his after life? I think it is possible from the study of his letters and other published writings to give an answer to this very interesting question.

In November 1832, Darwin returned to Monte
Video, from a long journey in the interior of the
South American Continent, bringing with him many
zoological specimens and a great quantity of fossil
bones, teeth and scales, dug out by him with infinite
toil from the red mud of the Pampas—these fossils
evidently belonging to the geological period that
immediately preceded that of the existing creation.
The living animals represented in his collection were
all obviously very distinct from those of Europe—
consisting of curious sloths, anteaters, and arma-
dilloes—the so-called 'Edentata' of naturalists.
And when young Darwin came to examine and
compare his *fossil* bones, teeth and scales he found
that they too must have belonged to animals
(megatherium, mylodon, glyptodon, etc.) quite distinct from but of strikingly similar structure to those
now living in South America. What could be the
meaning of this wonderful analogy? If Cuvier and
his fellow Catastrophists were correct in their view
that, at each 'revolution' taking place on the earth's
surface, the whole batch of plants and animals was
swept out of existence, and the world was re-stocked
with a 'new creation,' why should the brand-new
forms, at any particular locality, have such a 'ghost-
like' resemblance to those that had gone before? It
is interesting to note that, just at the same time,
a similar discovery was made with respect to Australia.

In caves in that country, a number of bones were
found which, though evidently belonging to 'extinct'
animals, yet must have belonged to forms resembling
the kangaroos and other 'pouched animals' (mar-
supials) now so distinctive of that continent.  But of
this fact Darwin was not aware until after his return
to England in 1836.

Among the objects sent from home, which awaited
Darwin on his return to Monte Video, was the second
volume of Lyell's *Principles,* then newly published ;
this book, while rejecting Lamarckism, was crowded
with facts and observations concerning variation,
hybridism, the struggle for existence, and many other
questions bearing on the great problem of the origin
of species.  I think there can be no doubt that from
this time Darwin came to regard the question of
species with an interest he had never felt before.

It is of course not suggested that, at this early
date, Darwin had formed any definite ideas as to the
*mode* in which new species might possibly arise from
pre-existing ones or even that he had been converted
to a belief in evolution.  Indeed in 1877 he wrote
'When I was on board the *Beagle* I believed in the
permanence of species' yet he adds 'but as far as
I can remember *vague doubts* occasionally flitted
across my mind.'  Such 'vague doubts' could scarcely
have failed to have arisen when, as happened during
all his journeys from north to south of the South

American Continent, he found the same curious correspondence between existing and late fossil forms of life again and again illustrated.

But towards the end of the voyage, an even stronger element of doubt as to the immutability of species was awakened in his mind. When he came to study the forms of life existing in the Galapagos Islands, off the west coast of South America, he was startled by the discovery of the following facts. Each small island had its own 'fauna' or assemblage of animals—this being very strikingly shown in the case of the reptiles and birds. And yet, though the *species* were different, there was obviously a very wonderful 'family likeness' to one another between the forms in the several islands and between them all and the animals living in the adjoining portion of the continent. Surely this could not be accidental, but must indicate relationships due to descent from common ancestors !

Charles Darwin returned to England in 1836, and at once made the acquaintance of Lyell. He says in one place, 'I saw a great deal of Lyell' and in another that 'I saw more of Lyell than of any other man, both before and after my marriage.' In one of his letters he writes, 'You cannot conceive anything more thoroughly good natured than the heart-and-soul manner in which he put himself in my place and thought what would be best to do[105].' For two

years Darwin was comparatively free from the
distressing malady which clouded so much of his
after life.  And, during that time, he engaged very
heartily with Lyell in those combats at the Geological
Society (of which he had become one of the Secretaries) in which their joint views concerning the truth
of continuity or evolution in the inorganic world
were defended against the attacks of the militant
catastrophists.  Darwin, however, did not act on the
defensive alone, but brought forward a number of
papers strongly supporting his new friend's views.

There can be little doubt that, while thus engaged, and in constant friendly intercourse with
Lyell, Darwin must have felt—like other earnest
thinkers on geology at that day—that the principles
they were advocating of ' continuity ' in the inorganic
world must be equally applicable to the organic
world—and thus that the question of evolution
would acquire a new interest for him.

But it was undoubtedly the revision of the notes
made on board the *Beagle*, and the study of the
specimens which had been sent home by him from
time to time, that produced the great determining
influence on Darwin's career.  All through the
voyage he had endeavoured, with as much literary
skill as he could command, to record with accuracy
the observations he made, and the conclusions to
which, on careful reflection, they seemed to point.

And on his return to England, these patiently written
journals were revised and prepared for publication
forming that charming work *A Naturalist's Voyage.
Journal of Researches into the Natural History and
Geology of the Countries visited during the Voyage
of H.M.S. 'Beagle' round the world.*

As Darwin, with the specimens before him, revised
his notes, and reconsidered the impressions made on
his mind, the 'vague doubts' he had entertained,
from time to time, concerning the immutability of
species, would come back to him with new force and
cumulative effect. 'I then saw,' he says, 'how many facts
indicated the common descent of species,' and further,
'It occurred to me in 1837, that something might
perhaps be made out on this question by patiently
accumulating and reflecting on all sorts of facts
which could possibly have any bearing on it.' In
July of that year, he opened his first note-book on
the subject[106]—the note-books being soon replaced by
a series of portfolios, in which extracts from the
various works he read, facts obtained by correspondence, the records of experiments and observation,
and ideas suggested by constant meditation were
slowly accumulated for twenty years. Mr Francis
Darwin has published a series of extracts from the
note-book of 1837, which amply prove that by this
time Charles Darwin had become 'a convinced
evolutionist[107].'

Fifteen months after this 'systematic enquiry'
began, Darwin happened to read the celebrated work
of Malthus *On Population,* for amusement, and this
served as a spark falling on a long prepared train
of thought. The idea that as animals and plants
multiply in geometrical progression, while the
supplies of food and space to be occupied remain
nearly constant, and that this must lead to a 'struggle
for existence' of the most desperate kind, was by no
means new to Darwin, for the elder De Candolle,
Lyell and others had enlarged upon it ; yet the facts
with regard to the human race, so strikingly presented by Malthus, brought the whole question with
such vividness before him that the idea of 'Natural
Selection' flashed upon Darwin's mind. This hypothesis cannot be better or more succinctly stated
than in Huxley's words.

'All *species* have been produced by the development of
*varieties* from common stocks : by the conversion of these, first
into *permanent races* and then into *new species,* by the process
of *natural selection,* which process is essentially identical with
that artificial selection by which man has originated the races of
domestic animals—the *struggle for existence* taking the place of
man, and exerting, in the case of natural selection, that selective
action which he performs in artificial selection[108].'

With characteristic caution, Darwin determined
not to write down 'even the briefest sketch' of this
hypothesis, that had so suddenly presented itself to

his mind.  His habit of thought was always to give
the fullest consideration and weight to any possible
objection that presented itself to his own mind or
could be suggested to him by others.  Though he was
satisfied as to the truth and importance of the principle
of natural selection, there is evidence that for some
years he was oppressed by difficulties, which I think
would have seemed greater to him than to anyone
else.  In my conversations with Darwin, in after
years, it always struck me that he attached an
exaggerated importance to the merest suggestion of
a view opposed to that he was himself inclined to
adopt ; indeed I sometimes almost feared to indicate
a *possible* different point of view to his own, for fear
of receiving such an answer as 'What a very striking
objection, how stupid of me not to see it before, I
must really reconsider the whole subject.'

While a divinity student at Cambridge, Darwin
had been much struck with the logical form of the
works both of Euclid and of Paley.  The rooms of
the latter he seems to have actually occupied at
Christ's College and the works of the great divine
were so diligently studied that their deep influence
remained with him in after life[109].

I think it must have been the remembrance of
the arguments of Paley on the 'proofs of design' in
Nature, that seem in after life to have haunted
Darwin so that for long he failed to recognise fully

that the principle of natural selection accounted not
only for the *adaptation* of an organism to its environment, but at the same time explains that *divergence*,
which must have taken place in species in order to
give rise to their wonderfully varied characters.

It was not till long after he came to Down in
1842, he tells us in his autobiography, that his mind
freed itself from this objection.   He says :—

'I can remember the very spot in the road, whilst in my
carriage, when to my joy the solution occurred to me,'

and he compares the relief to his mind as resembling
the effect produced by 'Columbus and his egg[110].'
Some may think the 'solution' of Columbus was
itself not a very satisfactory one ; and I am inclined
to regard the difficulties of which Darwin records so
sudden and dramatic a removal as more imaginary
than real !

There can be no doubt that, as pointed out by the
late Professor Alfred Newton[111], there was among
naturalists during the second quarter of the nineteenth century a feeling of dissatisfaction with
respect to current ideas concerning the origin of
species, accompanied in many cases with one of
expectation that a solution might soon be found.
Others, however, despairingly regarded it as 'the
mystery of mysteries' for which it was hopeless to
attempt to find a key.   There was, however, one
man, who simultaneously with Darwin was meditating

earnestly on the problem and who eventually reached the same goal.

Alfred Russel Wallace was born thirteen years after Darwin, and a quarter of a century after Lyell. He did not possess the moderate income that permits of entire devotion to scientific research—an advantage, the importance of which in their own cases, both Lyell and Darwin were always so ready to acknowledge. Wallace, after working for a time as a land-surveyor and then as a teacher, at the age of 26 set off with another naturalist, H. W. Bates, on a collecting tour in South America—hoping by the sale of specimens to cover the expenses of travel. Like Lyell and Darwin, he was an enthusiastic entomologist, and had conceived the same passion for travel. He had, as we have already seen, been deeply impressed by reading the *Principles of Geology*, and after spending four years in South America undertook a second collecting tour, which lasted twice that time, in the Malay Archipelago.

Before leaving England in 1848, Wallace had read and been impressed by reading the *Vestiges of Creation*, and there can be no doubt that from that period the question of evolution was always more or less distinctly present in his mind. While in Sarawak in the wet season, he tells us, ' I was quite alone with one Malay boy as cook, and during the evenings and wet days I had nothing to do but to look over my books and ponder over the problem which was rarely

absent from my thoughts.' He goes on to say that
by 'combining the ideas he had derived from his
books that treated of the distribution of plants and
animals with those he obtained from the great work
of Lyell' he thought 'some valuable conclusions
might be reached[112].' Thus originated the very
remarkable paper, *On the Law which has regulated
the Introduction of New Species*, the main conclusion
of which was as follows : 'Every species has come into
existence coincident both in space and time with a
pre-existing closely allied species.' As Wallace has
himself said, 'This clearly pointed to some kind of
evolution...but the *how* was still a secret.'

This essay was published in the *Annals and
Magazine of Natural History* in September 1855  It
attracted much attention from Lyell and Darwin and
later from Huxley.  One important result of it was
that Darwin and Wallace entered into friendly
correspondence.  But although Darwin in his letters
to Wallace informed him that he had been engaged
for a long time in collecting facts which bore on the
question of the origin of species, he gave no hint of
the theory of natural selection he had conceived
seventeen years before—indeed his friends Lyell and
Hooker appear at that time to have been the only
persons, outside his family circle, whom he had taken
into his confidence.

In the spring of 1858, Wallace was at Ternate in
the island of Celebes, where he lay sick with fever,

and as his thoughts wandered to the ever-present
problem of species, there suddenly recurred to his
memory the writings of Malthus, which he had read
twelve years before. Then and there, 'in a sudden
flash of insight' the idea of natural selection presented itself to his mind, and after a few hours'
thought the chief points were written down, and
within a week the matter was 'copied on thin letter-
paper' and sent to Darwin by the next post, with a
letter to the following effect[113]. Wallace stated that
the idea seemed new to himself and he asked Darwin,
if he also thought it new, to show it to Lyell, who
had taken so much interest in his former paper.
Little did Wallace think, in the absence of all
knowledge on his part of Darwin's own conclusions,
what stir would be made by his paper when it arrived
in England!

Wallace's essay was entitled *On the Tendency of
Varieties to depart indefinitely from the Original
Type,* and it is a singularly lucid and striking
presentment, in small compass, of the theory of
Natural Selection.

Had these two men been of less noble and
generous nature, the history of science might have
been dishonoured by a painful discussion on a
question of priority. Fortunately we are not called
upon for anything like a judicial investigation of
rival claims ; for Darwin as soon as he read the essay
saw that—as Lyell had often warned him might be

the case—he was completely forestalled in the
publication of his theory.  The letter and paper
arrived at a sad time for Darwin—he was at the
moment very ill, there was 'scarlet fever raging in
his family, to which an infant son had succumbed
on the previous day, and a daughter was ill with
diphtheria[114].'  Darwin at once wrote hurriedly to
Lyell enclosing the essay and saying :

'I never saw a more striking coincidence; if Wallace had my
MS. sketch written out in 1842, he could not have made a better
short abstract!  Even his terms now stand as heads of my
chapters.  Please return me the MS., which he does not say he
wishes me to publish, but I shall, of course, at once write and
offer to send to any journal.  So all my originality, whatever it
may amount to, will be smashed, though my book, if it ever have
any value, will not be deteriorated, as all the labour consists in
the application of the theory.  I hope you will approve of
Wallace's sketch, that I may tell him what to say[115].'

And Wallace—what was the line taken by him in
the unfortunate complication that had thus arisen ?
From the very first his action was all that is generous
and noble.  Not only did he, from the first, entirely
acquiesce in the course taken by Lyell and Hooker,
but, writing in 1870, when the fame of Darwin's work
had reached its full height, he said :—

'I have felt all my life and I still feel, the most sincere
satisfaction that Mr Darwin had been at work long before me,
and that it was not left for me to attempt to write *The Origin of
Species*.  I have long since measured my own strength and

know well that it would be quite unequal to that task.  For
abler men than myself may confess, that they have not that
untiring patience in accumulating, and that wonderful skill in
using, large masses of facts of the most varied kind,—that wide
and accurate physiological knowledge,—that acuteness in devising
and skill in carrying out experiments,—and that admirable style
of composition, at once clear, persuasive and judicial,—qualities
which in their harmonious combination mark out Mr Darwin as
the man, perhaps of all men now living, best fitted for the great
work he has undertaken and accomplished[116].'

And fifty years after the joint publication of the
theory of Natural Selection to the Linnean Society
he said:

'*I* was then (as often since) the "young man in a hurry," *he*'
(Darwin) 'the painstaking and patient student, seeking ever the
full demonstration of the truth he had discovered, rather than to
achieve immediate personal fame[117].'

And when he referred to the respective shares of
Darwin and himself to the credit of having brought
forward the theory of natural selection, he actually
suggests as a fair proportion '*twenty years to one
week*'—those being the periods each had devoted to
the subject[118]!

Never surely was such a noble example of
personal abnegation! We admire the generosity,
though we cannot accept the estimate, for do we not
know that, for at least half the period of Darwin's
patient quest, Wallace had spent in deeply pondering
upon the same great question?

# CHAPTER X

## THE ORIGIN OF SPECIES

In the preceding chapter I have endeavoured to
show how the hypothesis of Natural Selection
originated in the minds of its authors, and must
now invite attention to the way in which it was
introduced to the world. What has been said earlier
with respect to the labours and writings of Hutton,
Scrope and Lyell may serve to indicate the great
importance of the *manner* of presentment of new
ideas—the logical force and literary skill with which
they are brought to the notice of scientific contemporaries and the world at large.

There are some striking passages in Darwin's
naive 'autobiography and letters' which indicate the
beginnings of his ambition for literary distinction.
It must always be borne in mind in reading this
autobiography, however, that it was not intended by
Darwin for publication, but only for the amusement
of the members of his own family. But the charming
and unsophisticated self-revelations in it will always
be a source of delight to the world.

8—2

When making his first original observations among
the volcanic cones and craters of St Jago in the
Cape-de-Verde Islands, he says 'It then first dawned
on me that I might perhaps write a book on the
geology of the different countries visited, and this
made me thrill with delight [119].'   He tells us concerning his regular occupations on board the *Beagle*, that
'during some part of the day, I wrote my Journal
and took much pains in describing carefully and
vividly all that I had seen: and this was good
practice [120].'

   'Later in the voyage' he says 'FitzRoy' (the
Captain of the *Beagle*) 'asked me to read some of my
Journal and declared it would be worth publishing,
so here was a second book in prospect [121] !'

   Darwin's first published writings were the extracts
from his letters which Henslow read to the Philosophical Society of Cambridge, and those which
Sedgwick submitted to the Geological Society.   At
Ascension, on the voyage home, a letter from
Darwin's sisters had informed him of the com-
mendation with which Sedgwick had spoken to his
father of these papers, and he wrote fifty years
afterwards: 'After reading this letter, I clambered
over the mountains of Ascension with a bounding
step, and made the volcanic rocks ring under my
geological hammer.'   When in 1839 his charming
*Journal of Researches* was published he records that

'The success of this my first literary child always
tickles my vanity more than that of any of my other
books[122].'

As a matter of fact, no one could possibly be
more diffident and modest about his actual literary
performances than was Charles Darwin.  I have heard
him again and again express a wish that he possessed
'dear old Lyell's literary skill'; and he often spoke
with the greatest enthusiasm of the 'clearness and
force of Huxley's style.'  On one occasion he mentioned to me, with something like sadness in his
voice, that it had been asserted 'there was a want of
connection and continuity in the written arguments,'
and he told me that, while engaged on the *Origin*,
he had seldom been able to write, without interruption from pain, for more than twenty minutes at
a time !

Charles Darwin never spoke definitely to me
about the nature of the sufferings that he so
patiently endured.  On the occasion of my first visit
to him at Down he wrote me a letter (dated
August 25th, 1880) in which, after giving the most
minute and kindly directions concerning the journey,
he arranged that his dog-cart should bring me to the
house in time for a 1 o'clock lunch, telling me that to
catch a certain train for return, it would be necessary
to leave his house a little before 4 o'clock.  But he
added significantly :—

'But I am bound to tell you that I shall not be able to talk
with you or anyone else for this length of time, however much I
should like to do so—but you can read newspaper or take a stroll
during part of the time.'

His constant practice, whenever I visited him,
either at Down or at his brother's or daughter's house
in London, was to retire with me, after lunch, to a
room where we could 'talk geology' for about three
quarters of an hour. At the end of that time,
Mrs Darwin would come in smilingly, and though no
word was spoken by her, Darwin would at once rise
and beg me to read the newspaper for a time, or, if I
preferred it, to take a stroll in the garden; and after
urging me to stay 'if I could possibly spare the time,'
would go away, as I understood to lie down. On his
return, about half an hour later, the discussion would
be resumed where it had been left off, without further
remark.

Mr Francis Darwin has told us that the nature
and extent of his father's sufferings—so patiently
and uncomplainingly borne—were never fully known,
even to his own children, but only to the faithful
wife who devoted her whole life to the care
of his health. As is well known, Darwin seldom
visited at other houses, besides those of immediate
relatives, or the hydropathic establishment at which
he sought relief from his illness. But he was in the
habit of sometimes, when in London, calling upon

David Forbes the mineralogist (a younger brother of
Edward Forbes) then living in York Street, Portman
Square.  The bonds of union between Charles Darwin
and David Forbes were, first, that they had both
travelled extensively in South America, and secondly,
that both were greatly interested in methods of
preserving and making available for future reference
all notes and memoranda collected from various
sources.  David Forbes devoted to the purpose a
large room with the most elaborate system of pigeon-
holes, about which he told me that Darwin was
greatly excited.  He also mentioned to me that, on
one or more occasions, while Darwin was in his
house, pains of such a violent character had seized
him that he had been compelled to lie down for a time
and had occasioned his host the greatest alarm.

It must always therefore be remembered, in
reading Darwin's works, what were the sad conditions
under which they were produced.  It seems to be
doubtful to what extent his ill-health may be
regarded as the result of an almost fatal malady,
from which he suffered in South America, or as the
effect of the constant and prolonged sea-sickness of
which he was the victim during the five years' voyage.
But certain it is that his work was carried on under
no ordinary difficulties, and that it was only by the
exercise of the sternest resolution, in devoting every
moment of time that he was free from pain to his

tasks, that he was able to accomplish his great undertakings.

I do not think, however, that any unprejudiced reader will regard Darwin's literary work as standing in need of anything like an apology. He always aims—and I think succeeds—at conveying his meaning in simple and direct language ; and in all his works there is manifest that undercurrent of quiet enthusiasm, which was so strikingly displayed in his conversation. It was delightful to witness the keen enjoyment with which he heard of any new fact or observation bearing on the pursuits in which he was engaged, and his generous nature always led him to attach an exaggerated value to any discovery or suggestion which might be brought to his knowledge— and to appraise the work of others above his own.

The most striking proof of the excellence and value of Darwin's literary work is the fact that his numerous books have attained a circulation, in their original form, probably surpassing that of any other scientific writings ever produced—and that, in translations, they have appealed to a wider circle of readers than any previous naturalist has ever addressed !

We have seen that the idea of Natural Selection 'flashed on' Darwin's mind in October 1838, and although he was himself inclined to think that his *complete* satisfaction with it, as a solution of the problem of the origin of species, was delayed to a considerably later date, yet I believe that this was only the result of his over-cautious temperament, and we must accept the date named as being that of the real birth of the hypothesis.

At this early date, too, it is evident that Darwin conceived the idea that he might accomplish for the principle of evolution in the organic world, what Lyell had done, in the *Principles*, for the inorganic world. To cite his own words, 'after my return to England it appeared to me that by following the example of Lyell in Geology, and by collecting all facts which bore in any way on the variation of animals and plants under domestication and nature, some light might perhaps be thrown on the whole subject[123].' 'In June 1842,' he says, 'I first *allowed* myself' (how significant is the phrase!) 'the satisfaction of writing a brief abstract of my theory in pencil in 35 pages[124].'

For many years it was thought that this first sketch of Darwin's great work had been lost. But after the death of Mrs Darwin in 1896, when the house at Down was vacated, the interesting MS. was found 'hidden in a cupboard under the stairs which was not used for papers of any value but rather as an overflow of matters he did not wish to destroy[125].' By the pious care of his son, this interesting MS.— hurriedly written and sometimes almost illegible—

has been given to the world, and it proves how
completely Darwin had, at that early date, thought
out the main lines of his future *opus magnum*.

Darwin, however, had no idea of publishing his
theory to the world until he was able to support it
by a great mass of facts and observations. Lyell,
again and again, warned him of the danger which
he incurred of being forestalled by other workers;
while his brother Erasmus constantly said to him,
'You will find that some one will have been before
you[126]!'

The utmost that Darwin could be persuaded to
do, however, was to enlarge his sketch of 1842 into
one of 230 pages. This he did in the summer of
1844. His manner of procedure seems to have been
that, keeping to the same general arrangement of
the matter as he had adopted in his original sketch,
he elaborated the arguments and added illustrations.
Each of the 35 pages of the pencilled sketch, as it
was dealt with, had a vertical line drawn across it
and was thrown aside. While the 'pencilled sketch'
of 1842 was little better than a collection of memoranda, which, though intelligible to the writer at the
time, are sometimes difficult either to decipher or
to understand the meaning of, the expanded work
of 1844 was a much more connected and readable
document, which Darwin caused to be carefully
copied out. The work was done in the summer

months, while he was absent from home, and unable
therefore to refer to his abundant notes—Darwin
speaks of it, therefore, as 'done from memory.'

The two sketches, as Mr Francis Darwin points
out, were each divided into two distinct parts, though
this arrangement is not adopted in the *Origin of
Species*, as finally published.  Charles Darwin on many
occasions spoke of having adopted the *Principles of
Geology* as his model.  That work as we have seen
consisted of a first portion (eventually expanded from
one to two volumes), in which the general principles
were enunciated and illustrated, and a second portion
(forming the third volume), in which those principles
were applied to deciphering the history of the globe
in the past.  I think that Darwin's original intention
was to follow a similar plan ; the first part of his
work dealing with the evidences derived from the
study of variation, crossing, the struggle for existence, etc., and the second to the proofs that natural
selection had really operated as illustrated by the
geological record, by the facts of geographical distribution, and by many curious phenomena exhibited
by plants and animals.  Although this plan was
eventually abandoned—no doubt wisely—when the
*Origin* came to be written, we cannot but recognise
in it another illustration of the great influence
exercised by Lyell and his works on Darwin—an influence the latter was always so ready to acknowledge.

On the 5th July 1844, Darwin wrote a letter to
his wife in which he said, 'I have just finished my
sketch of my species theory.  If, as I believe, my
theory in time be accepted, even by one competent
judge, it will be a considerable step in science.'  He
goes on to request his wife, 'in case of my sudden
death' to devote £400 (or if found necessary £500)
to securing an editor and publishing the work.  As
editor he says 'Lyell would be the best, if he would
undertake it,' and later, 'Lyell, especially with the aid
of Hooker (and if any good zoological aid), would be
best of all.'  He then suggests other names from
which a choice might be made, but adds 'the editor
must be a geologist as well as naturalist.'  Fortunately
for the world Mrs Darwin was never called upon to
take action in accordance with the terms of this
affecting document[127].

It must be remembered that, at this time, Darwin
was hard at work on the three volumes of the
*Geology of the Beagle,* and on the second and revised
edition of his *Journal of Researches.*  This which he
considered his 'proper work' he stuck to closely,
whenever his health permitted.  He had hoped to
complete these books in three or four years, but
they actually occupied him for *ten,* owing to constant
interruptions from illness.  His occasional neglect of
this task, and indulgence in his 'species work,' as he
called it, was always spoken of at this time by

Darwin as 'idleness.' And when the geological and
narrative books were finished, Darwin took up the
systematic study of the Barnacles (*Cirripedia*), both
recent and fossil, and wrote two monumental works
on the subject. These occupied eight years, two out
of which he estimated were lost by interruptions
from illness. So absorbed was he in this work, that
his children regarded it as the *necessary occupation*
of a man,—and when a visitor in the house was seen
not to be so employed one of them enquired of their
mother, 'When does Mr —— do *his* Barnacles?'
Huxley has left on record his view that in devoting
so long a time to the study of the Barnacles Darwin
'never did a wiser thing,' for it brought him into
direct contact with the principles on which naturalists
found 'species[128].' And Hooker has expressed the
same opinion.

During these years of labour in geology and
zoology—interrupted only by the 'hours of idleness'
—devoted to 'the species question,' Darwin, though
leading at Down almost the life of a hermit, was
nevertheless in frequent communication with two or
three faithful friends who followed his labours with
the deepest interest. Cautious as was Darwin himself, he found in his life-long friend Lyell, a still more
doubting and critical spirit, and it is clear from what
Darwin says that he derived much help by laying
new ideas and suggestions before him. The year

before Darwin's death he wrote of Lyell, 'When I made a remark to him on Geology, he never rested till he saw the whole case clearly, and often made me see it more clearly than I had done before.'

Lyell's father was a botanist of considerable repute, the friend of Sir William Hooker and his distinguished son Dr (now Sir Joseph) Hooker. While Darwin was writing his *Journal of Researches*, he handed the proof-sheets to Lyell with permission to show them to his father, who was a man of great literary judgment. The elder Lyell, in turn, showed them to young Mr Hooker, who was then preparing to join Sir James Ross, in his celebrated Antarctic voyage with H.M. ships *Erebus* and *Terror*. Hooker was then working hard to take his doctor's degree before joining the expedition as surgeon, but he kept Darwin's proof-sheets under his pillow, so as to get opportunities of reading them 'between waking and rising.' Before leaving England, however, Hooker in 1839 casually met and was introduced to Darwin, and thus commenced a friendship which resulted in such inestimable benefits to science. Before sailing with the Antarctic expedition the young surgeon received from Charles Lyell, as a parting gift, 'a copy of Darwin's *Journal* complete'; and he tells us that the perusal stimulated in him 'an enthusiasm in the desire to travel and observe[129].'

On Hooker's return from the voyage in 1843,

a friendly letter from Darwin commenced that remarkable correspondence, which will always afford
the best means of judging of the development of
ideas in Darwin's mind.  Hooker's wide knowledge
of plants—especially of all questions concerning
their distribution—was of invaluable assistance to
Darwin, at a time when his attention was more
particularly absorbed by geology and zoology, while
botany had not as yet received much attention from
him.   Hooker's experience, gained in travel, his
sound judgment and balanced mind made him a
judicious adviser, while his caution and candour
fitted him to become a trenchant critic of new suggestions, scarcely inferior in that respect to Lyell.

Darwin does not appear to have made the
acquaintance of Huxley till a considerably later date ;
but we find the great comparative anatomist had in
1851 already become so deeply impressed by Darwin,
that he said in writing to a friend he 'might be
anything if he had good health[130].'  Huxley used to
visit Darwin at Down occasionally, and I have often
heard the latter speak of the instruction and pleasure
he enjoyed from their intercourse.

For many years of his life, Darwin used to come
to London and stay with his brother or daughter for
about a week at a time, and on these occasions—
which usually occurred about twice in the year I
believe—he would meet Lyell to 'talk Geology,'

Hooker for discussions on Botany, and Huxley for Zoology.

For twenty years Darwin had 'collected facts on a wholesale scale, more especially with respect to domesticated productions, by printed enquiries, by conversations with skilful breeders and gardeners, and by extensive reading.' 'When,' he added, 'I see the list of books of all kinds which I read and abstracted, including whole series of Journals and Transactions, I am surprised at my industry [131].' In September 1854 the Barnacle work was finished and 10,000 specimens sent out of the house and distributed, and then he devoted himself to arranging his 'huge pile of notes, to observing and experimenting in relation to the transmutation of species.'

It was early in 1856 when this work had been completed, that, again urged by Lyell, he actually commenced writing his book. It was planned as a work on a considerable scale and, if finished, would have reached dimensions three or four times as great as did eventually the *Origin of Species.* Working steadily and continuously he had got as far as Chapter X, completing more than one half the book, when as he says Wallace's letter and essay came 'like a bolt from the blue.'

Oppressed by illness, anxiety and perplexity, as we have seen that Darwin was at the time, he fortunately consented to leave matters—though with

great reluctance—in the hands of his friends Lyell
and Hooker. They took the wise course of reading
Wallace's paper at the Linnean Society on July 1st,
1858, at the same time giving extracts from Darwin's
memoir written in 1844, and the abstract of a letter
written by Darwin in 1857 to the distinguished
American botanist, Asa Gray. This solution of the
difficulty happily met with the complete approval of
Wallace ; and, as the result of the episode, Darwin
came to the conclusion that it would not be wise to
defer full publication of his views, until the extensive
work on which he was engaged could be finished, but
an 'abstract' of them must be prepared and issued
with as little delay as possible.

For a time there was hesitation, as Darwin's
correspondence with Lyell and Hooker shows, between the two plans of sending this 'abstract' to the
Linnean Society in a series of papers or of making
it an independent book. But Darwin entertained an
invincible dislike to submitting his various conclusions
to the judgment of the Council of a Society, and, in
the end, the preparation of the 'Abstract' in the
form of a book of moderate size, was decided on.
This was the origin of Darwin's great work.

The sickness at Down had led to the abandonment
of the house for a time, and, three weeks after the
reading of the joint paper at the Linnean Society,
we find Darwin temporarily established at Sandown,

in the Isle of Wight, where the writing of the *Origin of Species* was commenced. The work was resumed in September when the family returned to Down, and from that time was pressed forward with the greatest diligence.

For the first half of the book, the task before Darwin was to condense, into less than one half their dimensions, the chapters he had already written for the large work as originally projected. But for the second half of the book, he had to expand directly from the essay of 1844.

So closely did Darwin apply himself to the work, that, by the end of March 28th, 1859, he was able to write to Lyell telling him that he hoped to be ready to go to press early in May, and asking advice about publication : he says, 'My Abstract will be about five hundred pages of the size of your first edition of the *Elements of Geology.*' Lyell introduced Darwin to John Murray, who had issued all his own works, and the present representative of that publishing firm has placed on record a very interesting account of the ever thoughtful and considerate relations between Darwin and his publishers, which were maintained to the end [132].

The MS. of the book seems to have been practically finished early in May, and Darwin's health then broke down for a time, so completely that he had to retire to a hydropathic establishment.

By June 21st he was able to write to Lyell 'I am
working very hard, but get on slowly, for I find that
my corrections are terrifically heavy, and the work
most difficult to me. I have corrected 130 pages,
and the volume will be about 500. I have tried my
best to make it clear and striking, but very much
fear that I have failed ; so many discussions are and
must be very perplexing. *I have done my best.* If
you had all my materials, I am sure you would have
made a splendid book. I long to finish, for I am
certainly worn out[138].' On September 10th the last
proof was corrected and the preparation of the
index commenced. At the meeting of the British
Association in Aberdeen, Lyell made the important
announcement of the approaching publication of the
great work. On November 24th the book was issued,
1250 copies having been printed, and Darwin wrote
to Murray, 'I am infinitely pleased and proud at the
appearance of my child.' The edition was sold out
in a day, and was followed early in the next year
by the issue of 3000 copies ; and untold thousands
have since appeared.

The writing of such a work as the *Origin of
Species,* in so short a time—especially taking into
consideration the condition of its author's health—
was a most remarkable feat. It would, of course,
not have been possible but for the fact that Darwin's
mind was completely saturated with the subject, and

that he had command of such an enormous body
of methodically arranged notes.  He showed the
greatest anxiety to convince his scientific contemporaries, and at the same time to make his meaning
clear to the general reader.  With the former object,
both MS. and printed proofs were submitted to the
criticism of Lyell and Hooker ; and the latter end
was obtained by sending the MS. to a lady friend,
Miss G. Tollet—she, as Darwin says 'being an
excellent judge of style, is going to look out errors
for me.'  Finally the proofs of the book were
carefully read by Mrs Darwin herself.

The splendid success achieved by the work is
a matter of history.  Its clearness of statement and
candour in reasoning pleased the general public ;
critics without any profound knowledge of natural
history were beguiled into the opinion that they
*understood* the whole matter !  and, according to
their varying tastes, indulged in shallow objection
or slightly offensive patronage.  The fully-anticipated,
theological vituperation was of course not lacking,
but most of the 'replies' to Darwin's arguments
were 'lifted' from the book itself, in which objections
to his views were honestly stated and candidly considered by the author.

The best testimony to the profound and far-
reaching character of the scientific discussions of
the *Origin of Species* is found in the fact that both

Hooker and Huxley, in spite of their wide knowledge
and long intercourse with Darwin, found the work,
so condensed were its reasonings, a 'very hard book'
to read, one on which it was difficult to pronounce
a judgment till after several perusals !

It would be idle to speculate at the present day
whether the cause of Evolution would have been
better served by the publication, as Darwin at one
time proposed, of a 'Preliminary Essay,' like that of
1844, or by the great work, which had been commenced and half completed in 1858, rather than by
the 'abstract,' in which the theory of Natural Selection
was in the end presented to the world.   Probably
the more moderate dimensions of the *Origin of
Species* made it far better suited for the general
reader; while the condensation which was necessitated
did not in the end militate against its influence with
men of science.   It will I think be now generally
conceded that the great success of this grand work
was fully deserved.   A subject of such complexity as
that which it dealt with could only be adequately
discussed in a manner that would demand careful
attention and thought on the part of the reader ;
and Darwin's well-weighed words, carefully balanced
sentences, and guarded reservations are admirably
adapted to the accomplishment of the difficult task
he had undertaken.   The *Origin of Species* has been
read by the millions with pleasure, and, at the same

time, by the deepest thinkers of the age with
conviction.

It is scarcely possible to refer to the literary style
of Darwin's work without a reference to a misconception arising from that very candid analysis of his
characteristics which he wrote for the satisfaction of
his family, but which has happily been given to the
world by his son. In his early life Darwin was
exceedingly fond of music, and took such delight in
good literature, especially poetry, that when on his
journeys in South America he found himself able to
carry only one book with him, the work chosen was
the poems of Milton—the former student of his own
Christ's College, Cambridge. But towards the end
of his life, Darwin had sadly to confess that he found
that he had quite lost the capacity of enjoying either
music or the noblest works of literature.

Some have argued that Darwin's scientific labours
must have actually proved destructive to his artistic
and literary tastes, and have even gone so far as to
assert—in spite of numerous examples to the contrary—that there is a natural antithesis between the
mental conditions that respectively favour scientific
and artistic excellence.

But I think there is a very simple explanation of
the loss by Darwin of his powers of enjoyment of
music and poetry, a loss which he evidently greatly
deplored. His scientific undertaking was so gigantic,

and, at the same time, his health was so broken and precarious, that he felt his only chance of success lay in utilizing, for the tasks before him, every moment that he was free from acute suffering and retained any power of working.  Consequently, when the self-imposed task of each day was completed, he found himself in a state of mental collapse.  Now to appreciate the beauties of fine music or the work of a great writer certainly demands that the mind should be fresh and unjaded, whereas, at the only times Darwin had for relaxation, he was quite unfitted for these higher delights.  We are not surprised then to learn that he sought and found relief in listening to his wife's reading of some pleasant novel or in the nightly game of backgammon, as the only means of resting his wearied brain.

No one who had the privilege of conversing with Darwin in his later years can doubt of his having retained to the end the full possession of his refined tastes as well as his great mental powers.  His love for and sympathy with every movement tending to progress—especially in the scientific and educational world—his devotion to his friends, with no little indulgence of indignation for what he thought false or mean in others, these were his conspicuous characteristics, and they were combined with a gentle playfulness and sense of humour, which made him the most delightful and loveable of companions.

# CHAPTER XI

## THE INFLUENCE OF DARWIN'S WORKS

In two essays 'On the Coming of Age of the Origin of Species[134],' and 'On the Reception of the Origin of Species[135],' published in 1880 and 1887 respectively, Huxley has discussed the course of events following the publication of Darwin's great work, he having the advantage of being one of the chief actors in those events. There is a striking parallelism between the manner that the *Principles of Geology* had been received thirty years earlier, and the way that the *Origin of Species* was met, both by Darwin's scientific contemporaries and the reading public.

At the outset, as we have already intimated, Lyell and Darwin were equally fortunate, in that each found a critic, in one of the chief organs of public opinion, who was at the same time both competent and sympathetic. The story of the lucky accident by which this came about in Darwin's case has been told by Huxley himself[136].

'The *Origin* was sent to Mr Lucas, one of the staff of the *Times* writers at that time, in what was I suppose the ordinary

course of business.  Mr Lucas, though an excellent journalist,...
was as innocent of any knowledge of science as a babe, and
bewailed himself to an acquaintance on having to deal with such
a book.   Whereupon, he was recommended to ask me to get him
out of the difficulty, and he applied to me accordingly, explaining,
however, that it would be necessary for him formally to adopt
anything I might be disposed to write, by prefacing it with two
or three paragraphs of his own.'

'I was too anxious to seize upon the opportunity thus
offered of giving the book a fair chance with the multitudinous
readers of the *Times*, to make any difficulty about conditions;
and being then very full of the subject, I wrote the article faster,
I think, than I ever wrote anything in my life, and sent it to
Mr Lucas who duly prefixed his opening sentences [137].'

Many journalists, however, were less conscientious
than Mr Lucas, and most of the other early notices of
the book were pretty equally divided between undiscriminating praise of it as a novelty and foolish
reprobations of its 'wickedness.'

It was fortunate that Darwin followed the strong
advice given to him by Lyell, and did not attempt to
reply to the adverse criticisms; for the only effect of
these was to arouse curiosity and thus to increase the
circulation of the book.

Although Darwin had wisely avoided the danger
of exciting prejudice against his work by definitely
applying the theory of Natural Selection to the case
of man—simply remarking, in order to avoid the
charge of concealing his views, that 'light would be

thrown on the origin of man and his history'—yet
friends and foes alike at once drew what was the
necessary corollary from the theory. It is as amusing, as it is surprising at the present day, to recall
the storm of prejudice which was excited. At the
British Association Meeting at Oxford in 1860, after
an American professor had indignantly asked the
question, 'Are we a fortuitous concourse of atoms?'
as a comment on Darwin's views, Dr Samuel Wilber-
force, the Bishop of Oxford, ended a clever but
flippant attack on the *Origin* by enquiring of Huxley,
who was present as Darwin's champion, if it 'was
through his grandfather or his grandmother that he
claimed his descent from a monkey?'

Huxley made the famous and well-deserved re-
tort :—

'I asserted—and I repeat—that a man has no reason to be
ashamed of having an ape for his grandfather. If there were an
ancestor whom I should feel ashamed in recalling, it would rather
be a *man*—a man of restless and versatile intellect—who not
content with success in his own sphere of activity, plunges into
scientific questions with which he has no real acquaintance, only
to obscure them by an aimless rhetoric, and distract the attention
of his hearers from the real point at issue by eloquent digressions
and skilled appeals to religious prejudice[138].'

The violent attack on Darwin's views by the
once-famous Bishop of Oxford was outdone, a few
years later, by an even more absurd outburst on the

part of Benjamin Disraeli, who—after stigmatising
Darwinism as the question 'Is man an ape or an
angel?'—declared magniloquently to the episcopal
chairman, 'My Lord, I am on the side of the angels!'

But in spite of attacks like these and numerous
bitter pasquinades and comic cartoons—perhaps to
some extent in consequence of them—Darwin's views
became widely known and eagerly discussed, so that
the circulation of the *Origin of Species* went up by
leaps and bounds. Nevertheless, as Huxley said,
'years had to pass away before misrepresentation,
ridicule and denunciation, ceased to be the most
notable constituents of the multitudinous criticisms
of his work which poured from the press.'

Among his contemporary men of science Darwin
could at first count few converts. Hooker, whose
candid and valuable criticisms of his friend's work
had been continued up to the very end during its
composition, did an eminent service to the cause
of Evolution by publishing, almost simultaneously
with the *Origin of Species*, his splendid memoir on
*The Flora of Australia, its Origin, Affinities, and
Distribution*, in which similar views were, not ob-
scurely, indicated. Of Lyell, Darwin's other friend
and counsellor, Huxley justly says:

'Lyell, up to that time a pillar of the antitransmutationists
(who regarded him, ever afterwards, as Pallas Athene may have
looked at Dian, after the Endymion affair), declared himself

a Darwinian, though not without putting in a serious *caveat*. Nevertheless, he was a tower of strength and his courageous stand for truth as against consistency, did him infinite honour[139].'

Huxley himself accepted the theory of Natural Selection—but not without some important reservations—these, however, did not prevent him from becoming its most ardent and successful champion. Darwin used to acknowledge Huxley's great service to him in undertaking the defence of the theory— a defence which his own hatred of controversy and the state of his health made him unwilling to undertake—by laughingly calling him 'my general agent!' while Huxley himself in replying to the critics, declared that he was 'Darwin's bulldog.'

Although, at first, Darwin was able to enumerate less than a dozen naturalists who were prepared to accept his views, while influential leaders of thought in science—like Richard Owen in this country and Louis Agassiz in America—were bitterly opposed to them, the theory gradually obtained supporters especially among the younger cultivators of botany, zoology and geology.

It is evident that Darwin for some time regarded his 'abstract,' as he called the *Origin of Species*, as only a temporary expedient—one to be superseded by the publication of the much more extended work, designed and commenced long before. Although the *Origin* was only published late in November 1859,

and he was called upon immediately to prepare a
second edition, we find that on January 1st, 1860,
Darwin began to arrange his materials for dealing with
the first great division of his subject, 'the variation
of animals and plants under domestication.' So
numerous and important were his notes and records
of experiments, however, that he soon found that to
expand the whole of the 'abstract,' on the same scale,
would be an impossible task for any one man, however
able and diligent. Unwilling that the results of
some of his special researches should be lost, he
wisely determined to issue them as separate books.
The first of these to appear was that on the *Fertilisation of Orchids*, a beautiful illustration of the
relation of insects to flowers in producing crossing.
He had been more than twenty years working and
experimenting on this subject, his interest in it having
been quickened by having read an almost forgotten
book of the botanist Sprengel. Almost at the same
time, and in following years, he wrote papers for the
Linnean Society on dimorphic and trimorphic forms
of flowers, and their bearing on the question of cross-
fertilisation. These papers were the foundation of
his well-known work, *The Different Forms of Flowers
on Plants of the same Species*. In the same way,
a paper read in 1864 to the Linnean Society was
subsequently expanded into *The Movements and
Habits of Climbing Plants*.

Owing to delays caused by the preparation and
publication of these books and frequent interruptions
from sickness, the work on variation did not appear
till 1868.  It was a very extensive piece of work in
two volumes, and, at its end, Darwin tentatively
propounded a hypothesis to account for the facts
of Heredity and Variation to which he gave the
name of 'pangenesis.'

Charles Darwin had reached the age of fifty, when
he wrote the *Origin of Species*.  At a very early
period in his career, he had resolved that he would
never start a new theory or revise an old one after
he was sixty; as he used laughingly to say, 'I have
seen too many of my friends make fools of themselves
by doing that.'  But as he approached this 'fatal age,'
one more subject of a theoretical and highly controversial nature remained to be dealt with, namely,
the question of the application of the theory of
natural selection to man, both as regards his physical
structure and his intellectual and moral characteristics.

Darwin tells us that in 1837 or '38, as soon as he
had become 'convinced that species were mutable
productions,' he 'could not avoid the belief that man
must come under the same law[140].'  From that time,
he began collecting facts bearing on the question.
As each of his children was born, he examined closely
the signs of dawning intelligence, and made notes of

the manner in which new sensations and passions were exhibited by them. His dog and other animals, for whom he always showed the greatest fondness, were closely watched with the object of noting correspondences between their mental and moral processes and their modes of exhibiting them and our own; while visits were made by him to the Zoological Gardens with the same object. By reading and correspondence also, an enormous mass of notes was collected, and on February 4th, 1868, having seen his great work on Variation under Domestication published, Darwin was able to make the entry in his diary, 'Began work on Man.'

As was usual with most of his works, Darwin underestimated the time required to complete it. Through all the years 1867—'68, '69 and '70 we find the entries in his diary 'working at *Descent of Man*,' and only early in the year 1871 was the book finished. His original plan of compressing his notes on the expression of the Emotions into a chapter at the end of the book proved to be impracticable, and the material was reserved for a new work. This work, *The Expression of the Emotions in Man and Animals*, was commenced directly the *Descent of Man* was out of hand, a rough copy was finished by April 27th, 1871, but the last proofs were not corrected till August 23rd, 1873.

In dealing with the question of the origin of the

human race, Darwin was led to propound his views concerning Sexual selection, the results of the preferences shown by males and females, respectively, not only among mankind, but in various other animals. It was with respect to some of the conclusions contained in this work that Wallace found himself unable to follow Darwin. Wallace maintained that while man's body could have been developed by Natural Selection, his intellectual and moral nature must have had a different origin. He also declined to adopt the theory of sexual selection, so far as it depends on preferences exhibited by females for beauty in the males. Wallace, however, in some respects has always been disposed to attach more importance to Natural Selection, as the greatest, if not the only factor in evolution, than Darwin himself.

It will be seen that although Darwin had in all probability thought out all his important theoretical conclusions before 1869, when he reached the 'fatal age,' yet, owing to various delays, the books, in which he embodied his views, had not all appeared till more than four years later.

Lyell, who was a convinced evolutionist before the publication of the *Principles of Geology*, as is shown by his letters,—and the fact is strongly insisted on both by Huxley and Haeckel[141],—was slow in coming into *complete* agreement with Darwin concerning the theory of Natural Selection. While he followed his

friend's investigations with the deepest interest, his
less sanguine nature led him often to despair of the
possibility of solving 'the mystery of mysteries.' As
Darwin wrote only a year before his own death, Lyell
'would advance all *possible* objections to my suggestions, and *even after these were exhausted* would
long *remain dubious*[142].' It is evident from the correspondence that Darwin was at times tempted to
become impatient with the friend, for whose advocacy
of his views he so deeply longed.  Fourteen years
after the publication of the *Origin of Species*, however, Lyell, in his *Antiquity of Man*, gave in his
adhesion to Darwin's theory but, even then, not in
the unqualified manner that the latter desired.  Yet
I have reason to know that some years before his
death, Lyell was able to assure his friend of his
*complete* agreement, and Darwin, six years after the
loss of his friend, wrote, 'His candour was highly
remarkable.  He exhibited this by becoming a convert to the Descent theory, though he had gained
much fame by opposing Lamarck's views, *and this
after he had grown old.*' Darwin adds that Lyell,
referring to the '*fatal* age' of sixty, said 'he hoped
that now he might be allowed to live[143]!'

When I first came into personal relations with
Darwin, after the death of Lyell in 1875, he was in
the habit of deprecating any idea of his writing on
theoretical questions.  He used to talk of 'playing

with plants and such things,' and undoubtedly derived
the greatest pleasure from his ingenious experimental
researches. The result of this 'play' in which Darwin
took such delight is seen in his books on the *Power
of Movement in Plants* and *Insectivorous Plants*;
full of the records of ingenious experiments and
patient observation.

It was a great relief to Darwin that his friend
Wallace was able in 1871 to undertake the preparation of a work on *The Geographical Distribution of
Animals*, for, on many points, the views held by
Wallace on this subject were more in accordance
with Darwin's own, than were those of Lyell and
Hooker. Nevertheless, on all questions connected
with the geographical distribution of plants, and the
causes by which they were brought about, Darwin
always expressed the fullest confidence in Hooker's
judgment, and the greatest satisfaction with his
results.

With regard to another great division of his work,
that dealing with the imperfection, but yet great
value, of the geological record, Darwin was always
anxious, when I met him, to learn of any new discoveries. But he felt that he had done all that was
possible in his outline of the subject in the *Origin*,
and that he must leave to palaeontologists all over
the world the filling in of these outlines. So great
was the delight with which he used to hear of new

discoveries in palaeontology, that I often recall our
conversations in these later days, when so many interesting forms of extinct animal and vegetable life—
veritable 'missing links'—are being discovered in all
parts of the globe, and wish that he could have known
of them.  They are indeed 'Facts for Darwin.'

Very happy indeed was Charles Darwin in the last
years of his useful life, in returning to his oldest 'love'
—geology.  In studying the action of earthworms he
found a geological study in which his rare powers of
ingenious experimentation could be employed with
profit.  His earliest published memoir had dealt with
the question, and for more than forty years with
dogged perseverance, he had laboured at it from time
to time.  It was delightful to watch his pleasure as
he examined what was going on in the flower-pots
full of mould in his study, and when his book was
published and favourably received, he rejoiced in
it as 'the child of his old age[144].'

Charles Darwin's death took place rather more
than twenty-two years after the publication of the
*Origin of Species*.  Before he passed away, he had
the satisfaction of knowing that the doctrine of evolution had come to be—mainly through his own great
efforts—the accepted creed of all naturalists and that
even for the world at large it had lost its imaginary
terrors.  As Huxley wrote a few days after our sad
loss, 'None have fought better, and none have been

more fortunate, than Charles Darwin.  He found a
great truth trodden underfoot, reviled by bigots, and
ridiculed by all the world ; he lived long enough to
see it, chiefly by his own efforts, irrefragably established in science, inseparably incorporated with the
common thoughts of men, and only hated and feared
by those who would revile, but dare not.   What shall
a man desire more than this[145] ? '

More than a quarter of a century has passed since
these words were written.  How during that period
the influence of Darwin's writings on human thought
has grown, in an accelerated ratio, will be seen by
anyone who will turn the pages of the memorial
volume—*Darwin and Modern Science*—published
fifty years after the *Origin of Species*.  Therein, not
only zoologists, botanists and geologists, but physicists,
chemists, anthropologists, psychologists, sociologists,
philologists, historians—and even politicians and theologians—are found testifying to the important part
which Darwin's great work has played, in revolution-
ising ideas and moulding thought in connexion with
all branches of knowledge and speculation.

# CHAPTER XII

## THE PLACE OF LYELL AND DARWIN IN HISTORY

FROM the account given in the foregoing pages, it will be seen that—without detracting from the merits of their predecessors or the value of the labours of their contemporaries—we must ascribe the work of establishing on a firm foundation of observation and reasoning the doctrine of evolution —both in the inorganic and the organic world—to the investigations and writings of Lyell and Darwin.

Lyell had to oppose the geologists of his day, who led by Buckland in this country and by Cuvier on the continent, were almost, without exception, hopelessly wedded to the doctrines of 'Catastrophism,' and bitterly antagonistic to all ideas savouring of continuity or evolution. And, in the same way, Darwin, at the outset, found himself face to face with a similarly hostile attitude, on the part of biologists, with respect to the mode of appearance of new species of plants and animals

While Darwin doubtless derived his inspiration,

and  much  valuable  aid,  from  the  *Principles  of
Geology*,  and  its  gifted  author,  yet  Lyell,  with  all  his
clearness  of  vision,  logical  faculty  and  literary  skill,
did  not  possess  the  strong  faith  and  resolute  courage—to  say  nothing  of  that  wonderful  tenacity  of
purpose  and  power  of  research  which  were  such
striking  characteristics  of  Darwin—which would have
enabled  him  to  do  for  the  organic  what  he  did  for
the  inorganic  world.   If  it  be  true,  as  Darwin  used
to  suggest,  that  the  *Origin  of  Species*  might  never
have  been  written  had  not  Lyell  first  produced  the
*Principles  of  Geology*,  I  believe  it  is  no  less  certain
that  the  crowning  of  Lyell's  great  edifice,  by  the
full  application  of  his  principles  to  the  world  of  living
beings,  could  only  have  been  accomplished  by  a  man
possessing,  in  unique  combination,  the  powers  of
observation,  experiment,  reasoning  and  criticism,
joined  to  unswerving  determination,  which  distinguished  Darwin.

Starting  from  Lyell's  most  advanced  post,  Darwin
boldly  advanced  into  regions  in  which  his  friend  was
unable  to  lead,  and  indeed  long  hesitated  to  follow.
Together,  for  nearly  forty  years,  the  two  men—
influencing  one  another 'as  iron  sharpeneth  iron'—
thought  and  communed  and  worked,  aided  at  all
times  by  the  wide  knowledge  and  judicious  criticism
of  the  sagacious  Hooker;  and  together  the  fame  of
these  men  will  go  down  to  posterity.

XII]OF EVOLUTION

There is a tendency, when a great man has passed from our midst, to estimate his merits and labours with undiscriminating, and often perhaps exaggerated, admiration ; and this excessive praise is too often followed by a reaction, as the result of which the idol of one generation becomes almost commonplace to the next. A still further period is required before the proper position of mental perspective is reached by us, and a just judgment can be formed of the man's real place in history. The reputations of both Lyell and Darwin have, I think, passed through both these two earlier phases of thought, and we may have arrived at the third stage.

There was one respect in which both Lyell and Darwin failed to satisfy many both of their contemporaries and successors. Lyell, like Hutton, always deprecated attempts to go back to a 'beginning,' while Darwin, who strongly supported Lyell in his geological views, was equally averse to speculations concerning the 'origin of life on the globe.' Scrope[146], and also Huxley[147] in his earlier days, held the opinion that it was legitimate to assume or imagine a beginning, from which, with ever diminishing energy, the existing 'comparatively quiet conditions,' thought to characterise the present order of the world, would be reached. Both Lyell and Darwin insisted that geology is a historical science, and must be treated as such quite distinct from Cosmogony. And in the end, Huxley accepted the same view[148]. 'Geology,' he asserted, 'is as much a historical science as archaeology.'

The sober historian has always had to contend against the traditional belief that 'there were giants on the earth in those days!' The love of the marvellous has always led to the ascription of past events to the work of demigods who were not of like powers and passions with ourselves. Hence the invention of those 'catastrophies'—in which the reputations of deities as well as of men and women have often suffered. It is the same tendency in the human mind which makes it so difficult to conceive of all the changes in the earth's surface-features and its inhabitants being due to similar operations to those still going on around us.

Lyell's views have constantly been misrepresented by the belief being ascribed to him that 'the forces operating on the globe have never acted with greater intensity than at the present day.' But his real position in this matter was a frankly 'agnostic' one. 'Bring me evidence,' he would have said, 'that changes have taken place on the globe, which cannot be accounted for by agencies still at work *when operating through sufficiently long periods of time*, and I will abandon my position.' But such evidence was not forthcoming in his day, and I do not think has ever been discovered since. Professor Sollas has very justly said, 'Geology has no need to return to the catastrophism of its youth ; in becoming evolutional it does not cease to remain essentially uniformitarian[149].'

Alfred Russel Wallace, who has always been as stout a defender of the views of Lyell as he has of those of Darwin, has given me his permission to quote from a letter he wrote me in 1888. After referring to what he regards as the weak and mistaken attacks on Lyell's teachings, 'which have of late years been so general among geologists,' he says :—

'I have always been surprised when men have advanced the view that volcanic action *must* have been greater when the earth was hotter, and entirely ignore the numerous indications that both subterranean and meteorological forces, even in Palaeozoic times, were of the same order of magnitude as they are now—and this I have always believed is what Lyell's teaching implies.'

I believe that Mr Wallace's expression, adopted from the mathematicians, 'the same order of magnitude,' would have met with Lyell's complete acquiescence. He was not so unwise as to suppose that, in the limited periods of human history, we must necessarily have had experience—even at Krakatoa or 'Skaptar Jokull'—of nature's greatest possible convulsions, but he fought tenaciously against any admission of 'cataclysms' that would belong to a totally different category to those of the present day.

Apart from theological objections, the most formidable obstacle to the reception of evolutionary ideas had always been the prejudice against the admission of vast duration of past geological time. It was unfortunate that, even when rational historical criticism had to a great extent neutralised the effect of Archbishop Usher's chronology, the mathematicians and physicists, assuming certain sources of heat in the earth and sun could have been the only possible ones, tried to set a limit to the time at the disposal of the geologist and biologist. Happily the discovery of radio-activity and the new sources of heat opened up by that discovery, have removed those objections, which were like a nightmare to both Geology and Biology.

Lyell used to relate the story of a man, who, from a condition of dire poverty, suddenly became the possessor of vast wealth, and when remonstrated with by friends on the inadequacy of a subscription he had offered, the poor fellow exclaimed sadly, 'Ah! you don't know how hard it is to get the chill of poverty out of one's bones.'

Geologists and biologists alike have long been the victims of this 'chill of poverty,' with respect to past time. So long as physicists insisted that one hundred millions, or forty millions, or even ten millions of years, must be the limit of geological time, it was not possible to avoid the conclusion stated by Lord Salisbury in 1894, 'Of course, if the mathematicians are right the biologists cannot have what they de-

mand[150].' But now geologists and biologists may
alike feel that the liberty with respect to *space*,
which is granted ungrudgingly to the astronomer, is
no longer withheld from them in regard to *time*. We
can say with old Lamarck :—

'For Nature, Time is nothing. It is never a difficulty, she
always has it at her disposal; and it is for her the means by which
she has accomplished the greatest as well as the least results.
For all the evolution of the earth and of living beings, Nature
needs but three elements—Space, Time and Matter[151].'

Darwin, equally with Lyell, has suffered from a
reaction following on extravagant and uninformed
praise of his work. The fields in which he laboured
single-handed, have yielded to hundreds of workers
in many lands an abundant harvest. New doctrines
and improved methods of enquiry have arisen—
Mutationism, Mendelism, Weismannism, Neo-Lam-
arckism, Biometrics, Eugenics and what not—are
being diligently exploited. But all of these vigorous
growths have their real roots in Darwinism. If we
study Darwin's correspondence, and the successive
essays in which he embodied his views at different
periods, we shall find, variation by mutation (or *per
saltum*), the influence of environment, the question of
the inheritance of acquired characters and similar
problems were constantly present to Darwin's ever
open mind, his views upon them changing from time
to time, as fresh facts were gathered.

No one could sympathise more fully than would
Darwin, were he still with us, in these various depar-
tures. He was compelled, from want of evidence,
to regard variations as spontaneous, but would have
heartily welcomed every attempt to discover the laws
which govern them ; and equally would he have
delighted in researches directed to the investigation
of the determining factors, controlling conditions and
limits of inheritance. The man who so carefully
counted and weighed his seeds in botanical experiments, could not but rejoice in the refined mathematical
methods now being applied to biological problems.

Let us not 'in looking at the trees, lose sight of
the wood.' Underlying all the problems, some of
them very hotly discussed at the present day, there
is the great central principle of Natural Selection
—which if not the sole factor in evolution, is undoubtedly a very important and potent one. It is
only necessary to compare the present position of
the Natural History sciences with that which existed
immediately before the publication of the *Origin of
Species*, to realise the greatness of Darwin's achievement.

The fame of both Lyell and Darwin will endure,
and their names will remain as closely linked as were
the two men in their lives, the two devoted friends,
whose remains found a meet resting-place, almost
side by side, in the Abbey of Westminster. Very

touching indeed was it to witness the marks of
affection between these two great men ; an affection
which remained undiminished to the end.  Lyell was
twelve years senior to Darwin, and died seven years
before his friend.  During the last year of Lyell's
life, I spent the summer with him at his home in
Forfarshire.  How well do I recollect the keenness
with which—in spite of a near-sightedness that had
increased with age almost to blindness—he still
devoted himself to geological work.  The 264 note-
books, all carefully indexed, were in constant use,
and visits were made to all the haunts of his youth,
with the frequent pathetic appeal to me, 'You must
lend me your eyes.'  In spite of age and weakness,
he would insist on clambering up the steepest hills
to show me where he had found glacial markings,
and would eagerly listen to my report on them.  But
the *great* delight of those days was the arrival of
a letter from Darwin!  Lyell was the recipient of
many honours, and he declined many more, when he
feared that they might interfere with the work to
which he had devoted his life, but the distinction he
prized most of all was that conferred on him by his
lifelong friend, who used to address him as 'My dear
old Master,' and subscribe himself 'Your affectionate
pupil.'

   During the seven years that elapsed after the
death of Lyell, I saw Darwin from time to time, for

he loved to hear 'what was doing' in his 'favourite science.' On board the *Beagle*, before he had met the man whose life and work were to be so closely linked with his own, he was in the habit of specially treasuring up any 'facts that would interest Mr Lyell'; in middle life he declared that 'when seeing a thing never seen by Lyell, one yet saw it partially through his eyes[152]'; and never, I think, did we meet after the friend was gone, without the oft repeated query, 'What would Lyell have said to that?'

These reminiscences of the past, in which I have ventured to indulge, may not inappropriately conclude with a reference to the last interview I was privileged to have with him, who was 'the noblest Roman of them all!' On the occasion of his last visit to London, in December, 1881, Charles Darwin wrote asking me to take lunch with him at his daughter's house, and to have 'a little talk' on geology. Greatly was I surprised at the vigour which he showed on that afternoon, for, contrary to his usual practice, he did not interrupt the conversation to retire and rest for a time, though I suggested the desirability of his doing so, and offered to stay. His brightness and animation, which were perhaps a little forced, struck me as so unusual that I laughingly suggested that he was 'renewing his youth.' Then a slight shade passed over his countenance—but only for a moment—as he told me that he had 'received his warning.' The

attack, to which his son has alluded, as being the
prelude to the end[153], had occurred during this visit
to town ; and he intimated to me that he knew his
heart was seriously affected. Never shall I forget
how, seeing my concern, he insisted on accompanying
me to the door, and how, with the ever kindly smile
on his countenance, he held my hand in a prolonged
grasp, that I sadly felt might perhaps be the last.
And so it proved.

And now all the world is united in the conviction
which Darwin so modestly expressed concerning his
own career, 'I believe that I have acted rightly in
steadily following and devoting myself to science ! '

For has not that *devotion* resulted in a complete
reform of the Natural-History Sciences ! The doctrine
of the 'immutability of species'—like that of ' Catastrophism ' in the inorganic world—has been eliminated
from the Biological sciences by Darwin, through his
*steadily following* the clues found by him during his
South American travels; and continuity is now as
much the accepted creed of botanists and zoologists
as it is of geologists. As a result of the labours of
Darwin, new lines of thought have been opened out,
fresh fields of investigation discovered, and the
infinite variety among living things has acquired
a grander aspect and a special significance. Very
justly, then, has Darwin been universally acclaimed
as 'the Newton of Natural History.'

# NOTES

In the following references, L.L.L. indicates the "Life and Letters of Sir Charles Lyell" by Mrs K. Lyell (1881), D.L.L. the "Life and Letters of Charles Darwin" by F. Darwin (1887), M.L.D. "More Letters of Charles Darwin" edited by F. Darwin and A. C. Seward (1903), and H.C.E. Huxley's "Collected Essays."

1. The Darwin-Wallace Celebration, Linn. Soc. (1908), p. 10.
2. Darwin and Modern Science (1909), pp. 152–170.
3. Pope, Essay on Man, Ep. I. lines 111–2.
4. Genesis, Chap. xxx. verses 31–43.
5. Brit. Assoc. Rep. 1900 (Bradford), pp. 916–920.
6. *Ibid.* 1909 (Winnipeg), pp. 491–493.
7. L.L.L. Vol. I. p. 468.
8. Origin of Species, Chap. xv. end.
9. Milton, Paradise Lost, Bk. VII. lines 454–466.
10. Edinb. Rev. LXIX. (July 1839), pp. 446–465.
11. Principles of Geology, Vol. I. (1830), p. 61.
12. Zittel, Hist. of Geol. &c. Eng. transl. p. 72.
13. Quart. Rev. Vol. XLVIII. (March 1832), p. 126.
14. Brit. Assoc. Rep. 1866 (Nottingham).
15. H.C.E. Vol. VIII. p. 315.
16. *Ibid.* p. 190.
17. D.L.L. Vol. II. pp. 179–204.
18. H.C.E. Vol. V. p. 101.
19. D.L.L. Vol. II. p. 190.
20. Edinb. Rev. Vol. LXIX. (July 1839), p. 455 *note.*
21. 'Theory of the Earth,' Vol. II. p. 67.
22. L.L.L. Vol. I. p. 272.
23. Brit. Assoc. Rep. 1833 (Cambridge), pp. 365–414.
24. Outlines of the Geology of England and Wales, p. xliv.

25. Illustrations of the Huttonian Theory, p. iii.
26. Edinb. Rev. LXIX. (July 1839), p. 455 *note*.
27. *Ibid.*
28. Zittel, Hist. of Geol. &c. Eng. transl. p. 141.
29. Considerations on Volcanoes, &c. (1825), pp. iv–vi.
30. Volcanoes of Central France, 2nd Ed. (1858), p. vii.
31. See Quart. Rev. Vol. XXXVI. (Oct. 1827), pp. 437–485.
32. L.L.L. Vol. I. p. 46.
33. Principles of Geology, Vol. II. 2nd Ed.
34. L.L.L. Vol. II. pp. 47–8.
35. *Ibid.* Vol. I. p. 268.
36. Environs de Paris (1811), p. 56.
37. Trans. Geol. Soc. 2nd Ser. Vol. II. pp. 73–96.
38. See Mantell's Geology of the Isle of Wight and L.L.L. Vol. I. pp. 114–122.
39. Hist. of Geol. &c. Eng. transl. p. 188.
40. L.L.L. Vol. I. p. 173.
41. British Critic and Theological Review (1830), p. 7 of the review.
42. L.L.L. Vol. I. p. 177.
43. Preface to Vol. III. of the 'Principles' (1833), p. vii.
44. L.L.L. Vol. I. pp. 233–4.
45. Charles Lyell and Modern Geology (1898), p. 214.
46. Proc. Geol. Soc. Vol. I. p. 374.
47. L.L.L. Vol. I. p. 196.
48. *Ibid.* Vol. I. p. 197.
49. Proc. Geol. Soc. Vol. I. pp. 145–9.
50. L.L.L. Vol. I. p. 253.
51. *Ibid.* Vol. I. p. 234.
52. *Ibid.* Vol. I. p. 271.
53. *Ibid.* Vol. I. p. 270.
54. *Ibid.* Vol. I. p. 271.
55. Quart. Rev. Vol. XLIII. (Oct. 1830), pp. 411–469 and Vol. LIII. (Sept. 1835), pp. 406–448. Both these reviews are by Scrope. The Review of the 2nd Vol. of the 'Principles,' Q.R. Vol. XLVII. (March 1832), pp. 103–132 is by Whewell.

56. L.L.L. Vol. i. p. 270.
57. *Ibid.* Vol. i. pp. 260–1.
58. *Ibid.* Vol. i. p. 314.
59. *Ibid.* Vol. i. p. 165.
60. M.L.D. Vol. ii. p. 232 and D.L.L. Vol. ii. p. 190.
61. L.L.L. Vol. i. pp. 316–7.
62. Proc. Geol. Soc. Vol. i. pp. 302–3.
63. L.L.L. Vol. ii. p. 41.
64. See also D.L.L. Vol. i. pp. 72–3.
65. Nineteenth Century, Oct. 1895, and Controverted Questions in Geology (1895), pp. 1–18.
66. M.L.D. Vol. ii. p. 117.
67. D.L.L. Vol. i. pp. 337–8 and p. 342.
68. Origin of Species, Chap. x. See also Darwin and Modern Science, pp. 337–385.
69. D.L.L. Vol. i. pp. 341–2.
70. L.L.L. Vol. ii. p. 44.
71. D.L.L. Vol. i. p. 296.
72. *Ibid.* p. 72.
73. *Ibid.* p. 71.
74. A. R. Wallace, 'My Life, &c.' (1905), Vol. i. p. 433.
75. The Darwin-Wallace Celebration, Linn. Soc. (1908), p. 118.
76. L.L.L. Vol. ii. p. 459.
77. Report of lecture at Forrester's Hall.
78. H.C.E. Vol. viii. p. 312.
79. D.L.L. Vol. ii. p. 190.
80. L.L.L. Vol. ii. pp. 2, 3.
81. *Ibid.* Vol. ii. p. 36.
82. *Ibid.* Vol. ii. p. 5.
83. D.L.L. Vol. i. p. 94.
84. L.L.L. Vol. i. pp. 417–8.
85. H. F. Osborn, 'From the Greeks to Darwin' (1894), p. 165.
86. *Loc. cit.* pp. 467–469.
87. L.L.L. Vol. i. p. 168.
88. *Ibid.* Vol. ii. p. 365.

89. D.L.L. Vol. II. pp. 23, 29, 39.
90. *Ibid.* Vol. III. p. 15 (see also pp. 11–14).
91. 'Origin of Species,' 6th Ed. (1875), p. xiv.
92. 'Darwin and Modern Science,' p. 125.
93. 'Origin of Species,' 6th Ed. (1875), pp. xvi, xvii.
94. M.L.D. Vol. I. p. 3.
95. D.L.L. Vol. I. p. 41.
96. *Ibid.* Vol. I. p. 41.
97. *Ibid.* Vol. I. p. 52.
98. *Ibid.* Vol. I. p. 58.
99. *Ibid.* Vol. I. p. 58.
100. H.C.E. Vol. II. p. 271.
101. D.L.L. Vol. I. p. 73.
102. *Ibid.* Vol. I. p. 263.
103. *Ibid.* Vol. I. p. 38.
104. H.C.E. Vol. II. p. 20.
105. D.L.L. Vol. I. p. 275.
106. *Ibid.* Vol. I. p. 83.
107. *Ibid.* Vol. II. pp. 5–10.
108. H.C.E. Vol. II. p. 71.
109. D.L.L. Vol. I. p. 47.
110. *Ibid.* Vol. I. p. 84.
111. Macmillan's Magazine, Feb. 1888, p. 241.
112. My Life, &c. Vol. I. p. 355.
113. Darwin-Wallace Celebration, Linn. Soc. (1908), pp. 6–7.
114. *Ibid.* pp. 14–16.
115. D.L.L. Vol. II. pp. 116–7.
116. 'Contributions to the Theory of Natural Selection' (1871), Preface, pp. iv, v.
117. Darwin-Wallace Celebration, Linn. Soc. (1908), p. 7.
118. *Ibid.* p. 7.
119. D.L.L. Vol. I. p. 66.
120. *Ibid.* Vol. I. pp. 62–3.
121. *Ibid.* Vol. I. p. 66.
122. *Ibid.* Vol. I. p. 66.

123. D.L.L. Vol. I. p. 83.
124. *Ibid.* Vol. I. p. 84.
125. 'The Foundations of the Origin of Species' (1909), p. xv.
126. Letter to A. R. Wallace, Christ's Coll. Mag. Vol. xxiii. (1909), p. 229.
127. D.L.L. Vol. ii. pp. 16–18.
128. *Ibid.* Vol. I. p. 347.
129. D.L.L. Vol. ii. pp. 19–21.
130. Huxley's Life and Letters (1900), Vol. I. p. 94.
131. D.L.L. Vol. I. p. 83.
132. Science Progress, Vol. iii. (1908), pp. 537–542.
133. D.L.L. Vol. ii. p. 160.
134. H.C.E. Vol. ii. pp. 227–243.
135. D.L.L. Vol. ii. pp. 179–204.
136. *Ibid.* Vol. ii. p. 255.
137. The Review is republished in H.C.E. Vol. ii. pp. 1–21.
138. Huxley's Life and Letters, Vol. I. pp. 179–189.
139. D.L.L. Vol. ii. p. 185.
140. *Ibid.* Vol. I. p. 93.
141. See Haeckel's 'History of Creation.'
142. D.L.L. Vol. I. p. 71.
143. *Ibid.* Vol. I. p. 72.
144. D.L.L. Vol. I. p. 98; Vol. iii. pp. 217–218.
145. H.C.E. Vol. ii. p. 247.
146. Quart. Rev. xliii. pp. 464–467 and Vol. liii. pp. 446–448.
147. H.C.E. Vol. viii. p. 315.
148. H.C.E. Vol. v. p. 99.
149. The Age of the Earth and other Geological Studies, p. 322.
150. Brit. Assoc. Rep. 1894 (Oxford), p. 13.
151. 'Hydrogéologie,' p. 67.
152. M.L.D. Vol. ii. p. 117.
153. D.L.L. Vol. iii. p. 356.

# INDEX

Adaptation, in relation to divergence of species, Darwin's recognition of, 108, 109

Agriculturalists, ideas of creation, 5, 6

ARNOLD, MATTHEW, on Lucretius and Darwin, 3, 4

Auvergne, N. Desmarest on, 17; Scrope on, 35; visited by Lyell and Murchison, 56, 57; their memoir on, 58

'Beagle,' H.M.S., Darwin's voyage in, 98, 99; narrative of, 106

BONNEY, T. G., estimate of amount of Lyell's travels by, 56, 57

Botanical works of Darwin, 141

*British Critic*, Whewell's review of Lyell in, 53

BRODERIP, W. J., aid given to Lyell by, 65; Vol. II. of *Principles* dedicated to, 65

BROWN, ROBERT, assistance to Lyell by, 47

BUCKLAND, Dr, on infant Geological Society, 26; champion of 'Catastrophism' in England, 27; his eccentricity, 42-44; 'Equestrian Geology' of, 28; influence on Lyell, 34, 44; 2nd edition of Vol. I. of *Principles* dedicated to, 44; his opposition to Lyell, 71

Cambridge, Darwin at, 97, 98

CANDOLLE, A. P. DE, on struggle for existence, 107

Catastrophism, origin of idea of, 14, 15; defined, 22; origin of term, 22; connexion with orthodoxy, 21; championed by Buckland, Sedgwick &c., 27; by Cuvier, 31, 50, 102; opposition by Lyell and Darwin to, 105

Centres of Creation, Lyell's views on, 65

CHAMBERS, ROBERT, publishes *Vestiges of Creation*, 92; his reasons for anonymity, 93

Chemists, part played in early days of Geological Society by, 26

Christ's College, Cambridge, the home of Milton and Darwin, 13; of Paley, 108

CLODD, E., his *Pioneers of Evolution*, 16

Continuity, term for Evolution suggested by Grove, 23

CONYBEARE, W. D., advocacy of Catastrophism, 27; criticism of Hutton, 28; misconception of Hutton, 29; on formation of Thames Valley, 58; friendship with Lyell, 69

Creation, legends of, 5-7; use of term by Lyell and Darwin,

𝕮𝖆𝖒𝖇𝖗𝖎𝖉𝖌𝖊:

PRINTED BY JOHN CLAY, M.A.
AT THE UNIVERSITY PRESS

Printed in the United States
By Bookmasters